Beginner's Gu

Television

CW00632040

Beginner's Guides are available on the following subjects:

Amateur Radio
Audio
Basic Programming
Building Construction
Cameras
Central Heating
Colour Television
Computers
Digital Electronics
Domestic Plumbing
Electric Wiring
Electronics
Fabric Dyeing and
    Printing
Gemmology
Home Energy Saving
Integrated Circuits
Microprocessors
Photography
Processing and Printing
Radio
Spinning
Super 8 Film Making
Tape Recording
Technical Illustration
Technical Writing
Television
Transistors
Video
Weaving
Woodturning
Woodworking

Other books by Gordon J. King

Audio Equipment Tests
Colour Television Servicing
Newnes Colour TV Servicing Manuals
Radio Circuits Explained
Radio, Television and Audio Test Instruments
Servicing Radio, Hi-Fi and TV Equipment
Servicing with the Oscilloscope
The Audio Handbook
The Practical Aerial Handbook

Beginner's Guide to

Television

Gordon J. King

TEng (CEI), AMIERE, FSERT, FSCTE, FISTC, FIPRE

Revised by

E. Trundle

MSERT, MRTS

Newnes Technical Books

Newnes Technical Books
is an imprint of the Butterworth Group
which has principal offices in
London, Boston, Durban, Singapore, Sydney, Toronto, Wellington

Originally published 1958 by George Newnes as
 'A Beginner's Guide to Television'
Second edition 1959
Third edition 1963, reprinted 1965
Fourth edition (entirely rewritten) 1968 as
 'Beginner's Guide to Television'
Fifth edition 1972, reprinted 1975, 1978, 1979, 1981
Sixth edition 1983

British Library Cataloguing in Publication Data

King, Gordon
    Beginner's guide to Television – 6th ed.
    1. Television – Apparatus and supplies
    I. Title
    621.388   TK6630

    ISBN 0-408-01215-3

Photoset by Butterworths Litho Preparation Department
Printed in England by Butler & Tanner Ltd, Frome, Somerset

Preface

> What is now proved was once only imagin'd.
> William Blake, *The Marriage of Heaven and Hell*

The book first saw the light of day under the hand of the late F. J. Camm in 1958. It was completely rewritten in 1968 by Gordon J. King and updated by him in 1972. The intervening 10 years have seen many changes in television, and the introduction of new techniques and devices.

Teletext and Prestel are firmly established, and videotape machines, cordless remote control, and TV games have become commonplace in the home. Even in today's fast-changing world, there can be few fields which evolve as rapidly as television technology. Motorcar engines and gas boilers have changed over the years but would still be recognisable to a Rip van Winkle technician from 1968; bring back such a TV man from 1968 and see how much of a 1983 television receiver he would be able to recognise!

In preparing the sixth edition, I have found it necessary to rewrite most of the text and introduce a large number of new illustrations. Redundant information on the 405-line system, valve technology, and early forms of picture and camera tubes has been removed, and I have used the vacated space to enlarge on the 625-line system and particularly to expand the sections on the television receiver. Among the newly-introduced subjects are 'chip' technology, videodisc recording, and the brave new world of digital TV.

I have striven to maintain Gordon King's lucid and down-to-earth style, and to keep the text concise and readable. In a book of this sort, with such wide terms of reference, it is inevitable that some questions will not be fully answered, and I hope that the reader, with appetite whetted as it were, will go on to consult more advanced books. Television is a fascinating and rewarding subject to study – good reading!

E. Trundle

Contents

1

Principles

Television means 'seeing from afar', and the idea of sending images by electrical means is a very old one. It does not have to involve radio links, cathode-ray tubes, or high frequencies, although conventional television utilises a combination of all three. If we took a selenium photocell (a device whose resistance varies with the amount of light falling upon it) and arranged for it to feed a filament bulb via a battery or simple amplifier, we would have a system capable of transmitting changes in light intensity between two widely-separated points. It is obvious, however, that this is of no practical value because the photocell can only respond to the average light falling on the whole surface of its sensor, and even if we focussed an image on that surface, the receiving lightbulb would only indicate the mean brightness of the whole image.

Image dissection

If the photocell-and-bulb arrangement is carried a step further, it becomes possible to generate and display a complete picture. Imagine a matrix of photocells in a rectangular bank, say 100 × 100 devices. If these were connected to a similar matrix of lightbulbs, with each photocell linked via a battery or amplifier to its corresponding bulb, any image focussed on the photocell matrix would be accurately reproduced at the receiving end, and the more elements (photocells and bulbs) in the system, the greater the definition obtainable. This idea is quite workable, but uneconomic and unwieldy in practical terms because of the vast number (10 000 in the above example) of wires and batteries or amplifiers required.

Scanning
To overcome the problems associated with sending many streams of information simultaneously, the principle of scanning is used. The outputs from our 10 000 photocells could be connected to the studs of a vast rotary switch in such a way that, as the switch rotated, the output of the top left-hand cell was first connected to its rotor, then the next on the right, and so on until the end of the line of cells was reached. The switch would then read-out the outputs of the next row of cells down, and so on until the entire image had been scanned in exactly the same way as we read a page in a book.

The output from the switch rotor would consist of a series of pulses representing an orderly sequential readout of the intensity of light falling on each picture element. When the last element (at the bottom right-hand corner) had been scanned, the switch would have turned full circle, ready to begin again at the first element, at the top of the left-hand side.

If we could arrange an identical switch at the receiving end, with its 10 000 studs connected to the bulb matrix in the same order as at the sending end, and with its rotor fed from the rotor of the 'sending switch', we would have a television system. It would be essential that the switches rotate in exact synchronism with each other, or the reproduced image would be an unintelligible muddle.

The important thing is that we now have only a single link between the sender and the receiver. Provided the link can faithfully reproduce the rapid changes in voltage corresponding to the photocell pulses, and the scanning rate is fast enough, good pictures are assured.

This is a mechanical analogy, but a good one, since it illustrates the ideas of picture elements, scanning, and bandwidth. Many types of mechanical scanning systems were

2

used in experiments in the pioneer days of television, and J. L. Baird's system went so far as to appear, in 240-line form, in regular television transmissions in the 1930s. Mechanical scanning systems soon gave way to all-electronic television, however, and the 625-line system in use today for colour and monochrome transmission is exactly the same in princple as the 405-line system suggested by Campbell-Swinton as early as 1908 and brought to fruition by EMI in the mid-1930s. Bulb matrices still find application in data and picture presentation, mainly in advertising and news readouts, and many can be found in the Piccadilly area of London.

The human eye
All television is an illusion, and we see television pictures because our eyes are incapable of registering in its true form a bright, fast-moving pinpoint of light on the screen of the picture tube. Instead of perceiving the light as an actual spot, our eyes discern its movement on the screen as a series of thin, closely-spaced horizontal lines. This stems from a subjective process called *persistence of vision*.

Human vision is essentially a process in which our eyes translate the energy of light radiation focussed within them into a chemical activity which in turn sends nervous impulses to the brain, informing it about the scene on which the eyes are focused.

The main optical cells associated with this translation are called *rods* and *cones*. These work via the retina, which is a photosensitive film upon which the image of the scene is focused by the eye's lens. The rods respond to the brightness component of the scene, while the cones add the colour sensations.

Because of the relatively slow decay time (about 80 milliseconds) of the eye/brain response when the light stimulus is removed, fast, repetitive movement of a light or an object conveys the impression of continuity. This accounts for the illusion of cinematography. A spot of light moving back and forth recurrently along the same horizontal path on the screen of a picture tube appears as a continuous horizontal line when the repetition frequency exceeds about 15 Hz (15

3

cycles per second). Flicker is apparent at low frequencies but almost completely disappears at about 50 Hz.

Electronic scanning

The design of a television picture tube is such that an intensely bright pinpoint of light is produced in the centre of the screen. This is called the *scanning spot* and arises as a result of the electron beam in the tube striking the fluorescent coating of the screen. Two coil sets arranged at right angles to each other on the picture-tube neck are energised by currents of almost sawtooth waveform. The magnetic fields that accompany these currents in the coils subject the electron beam passing along the tube neck to vertical and horizontal forces. The beam is thus deflected; and since the electron beam is responsible for the creation of the scanning spot, this too is deflected both vertically and horizontally on the screen.

The tube neck coils are called *scanning coils*. There are two pairs of them, one pair for vertical deflection and the other pair for horizontal deflection.

Vertical deflection of the scanning spot is from the top to the bottom of the screen, and the British repetition rate is 50 Hz or thereabouts, originally to match the AC mains frequency. In America and other countries with a 60 Hz mains system, the vertical deflection rate is 60 Hz.

Horizontal deflection of the spot is from the left to the right of the screen (looking at it from the front of the tube) at a rate of 15 625 Hz on the 625-line standard.

Because the spot is deflected horizontally more swiftly than it is vertically, it traces on the screen a succession of horizontal lines, as shown in *Figure 1.1*. These are called *scanning lines*. The resulting rectangle of screen illumination, comprised by all the lines, is called the *raster*; it is upon this that the television picture is built.

The sawtooth nature of currents in the scanning coils causes the scanning spot linearly to deflect from top to bottom and from left to right on the screen, and after each

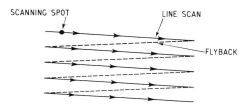

SCANNING SPOT   LINE SCAN

FLYBACK

Figure 1.1. The principles of scanning. Scanning coils deflect the spot horizontally more rapidly than vertically to give the effect of a succession of horizontal scanning lines

full deflection, the spot is made to return very rapidly to its starting position because of the rapid change in scanning-coil current at the end of each full deflection. A full deflection is called a *scanning stroke*, and the rapid return of the spot is called the *retrace stroke* or, often, the *flyback*.

Line standards
A television raster is composed of a certain number of scanning lines according to the line standard in use. The original 405-line standard adopted in Great Britain in 1937 is now obsolescent, and the UK standard, in common with that in most countries of the world, is 625 lines. There is also a 525-line standard, used in the USA, South America, and Japan.

Although a picture is said to have 625 lines, this is not strictly true. If it were possible to count the number of lines, it would be discovered that it is slightly less than the line standard. This is because some of the lines occur during the periods when there is no picture information being transmitted, known as the 'field blanking intervals'. A 625-line picture has about 575 active lines. The blanking periods between these lines contain synchronising signals, test waveforms, and teletext data, as we shall see in later chapters.

Frames and fields
Because the vertical repetition frequency is 50 Hz, it follows that a complete set of lines, forming the raster, must appear

5

on the screen every fiftieth of a second. However, we have seen that the line-repitition frequency is 15 625 Hz on the 625-line system. If we divide 15 625 by the vertical frequency of 50 Hz, we get an answer of 312½, and not 625 as might well be expected. This simple arithmetic suggests that for each vertical scan there is only half the number of lines of a complete picture. How, then, is the 625 standard created?

The answer is that each complete television picture, corresponding to one frame of a cine film, is made up of two vertical scans, each scan yielding half the number of lines of a complete picture. This means that a complete picture is produced every twenty-fifth of a second.

Each vertical scan of half the number of total picture lines is called a *field*; each composite picture, consisting of two fields, is called a *frame*. We thus get the term *field frequency*, which is the same as the vertical deflection already considered. The British field frequency is therefore 50 Hz, and that in America and some other countries 60 Hz.

As we have discovered, not all the lines of a field are active (picture-carrying) lines. About 25 per field are blanked on the 625/50 system.

Our persistence of vision is such that our eyes retain each field for a small period of time. The two fields are thus subjectively integrated, and we see a complete frame in spite of the fact that we are really looking at two fields occurring at intervals of one-fiftieth of a second.

Interlacing

The complete full-line picture or frame is seen because the scanning lines of one field interlace in the spaces between the lines of the partnering field, as shown in *Figure 1.2*.

The question now arises, why go to all the trouble of interlacing the lines of two fields to develop a whole picture when a full-line picture could surely be produced with a single vertical scan simply by doubling the line frequency? It is certainly true that a raster of 625 lines could be obtained by stepping up the line frequency to 31 250 Hz while retaining the original 50 Hz vertical frequency. The problem, however, is concerned with the horizontal definition of the picture,

6

which is influenced by the speed at which the spot traces out the scanning lines.

The more swiftly the spot moves on the horizontal scanning stroke, the greater the speed at which its brightness needs to change to 'write in' picture detail on the screen. The brightness change is controlled by the speed at which the level of the vision signals can change in the television system as a whole. Since this in turn is a factor of the system bandwidth, any increase in horizontal scanning speed demands an increase in bandwidth. Decreasing the bandwidth

Figure 1.2.   The scanning lines of the two vertical fields interlace

slows down the brightness change of the spot; the reason for this is considered later. As there is already a shortage of 'radio space' in the television bands, any scheme that increases the bandwidth requirement could never be used.

An alternative is to reduce the vertical frequency to 25 Hz and retain the original line speeds. True, this would eliminate the bandwidth demands of increased line speed, but the slowing down of the vertical scan would produce bad picture flicker, which is troublesome up to about 48 Hz.

Clearly, interlacing solves both these problems and allows the transmission and reception of good-quality pictures within a practical bandwidth at an effectively flicker-free vertical scanning frequency. Although only half the picture information is portrayed at a rate of 50 Hz by interlacing, this represents the flicker frequency in spite of the fact that full-line pictures are produced at the lower rate of 25 Hz.

It is important that the receiver is able to 'slot-in' the second field so that its scanning lines interleave between those of the first, as illustrated in *Figure 1.2*. If this does not happen and the lines become superimposed, vertical definition is halved.

Camera scanning
The image of a scene focused by a television camera is scanned by a focused electron beam just as the screen of the picture tube is scanned. The camera beam, however, does not result in a spot of light, but instead causes the camera to deliver information on that part of the image upon which it is impinging at any instant.

This information is given in the form of electrical impulses. Since the camera electron beam is being deflected both vertically and horizontally in exact synchronism with the picture-tube electron beam, the brightness of the scanning spot on the screen of the picture tube changes from instant to instant to match the brightness of that part of the image upon which the camera beam is impinging.

A microphone translates sound waves into electrical impulses, and a television camera does likewise to light waves. But as we have discovered from the photocells and bulbs, it is impossible to translate vision, as a complete scene, into one signal pattern. This can be done with sound because the signal waveform takes on the exact electrical character of the sound at any moment while it is being translated by the microphone. With vision, it is necessary to break down the scene into elements and to translate these by the camera to electrical impulses. Each picture element is translated sequentially so that a series or train of impulses makes up one line of picture information. On each line scan, therefore, there occurs a series of impulses which are electrically equivalent to the brightness and detail of the scene along each line.

This is where the scanning process comes in. The whole of the image of the scene is scanned at the camera, and electrical impulses are produced during the process of scanning. These impulses eventually arrive at the receiver, where they are used to change the brightness of the scanning spot on the picture-tube screen. Since this spot is always in the same relative position on the screen as the focused electron beam in the camera, the image 'seen' by the camera is built up on the picture-tube screen.

Definition

While the vertical definition of a television picture is governed by the number of scanning lines (i.e., the line standard), the horizontal definition is governed by the speed at which the brightness of the scanning spot can change and hence, as we have seen, by system bandwidth. It is possible to secure very high vertical definition, therefore, simply by increasing the number of lines, while the horizontal definition may be very poor. Apart from deleting the 'line effect', there is little point in increasing the vertical definition of a picture by increasing the number of lines beyond the value fixed by the horizontal definition and available bandwidth. Indeed, if the number of lines is increased without a corresponding increase in bandwidth, the overall definition of the picture will suffer.

Optimum overall definition occurs when the vertical and horizontal definitions are equal. This can be better appreciated in terms of picture elements.

Picture elements

Let us suppose that we have a 625-line picture geared to a raster whose height equals its width. We have seen that there are about 575 active lines in a 625-line picture, so if we multiply 575 by 575 we shall find how many picture elements exist in this square picture. The answer is 330 625.

In this simple illustration, it is assumed that the scanning spot has a diameter equal to the width of a scanning line, the latter being based on the height of the screen divided into 575 equal-width strips. Of course, in practice it is the diameter of the scanning spot which determines the width of the lines. A spot a bit too large would cause overlapping of the lines, while one too small would leave gaps between them. It is the job of the picture-tube maker to ensure that the spot produced by his tubes will give a scanning line of optimum width when the tubes are operating correctly. Bad circuit design or a fault in the set can make the spot size too small or too large.

Figure 1.3 represents a 10-line, square picture of very low definition. It is easier to appreciate the principles involved when the lines are only few; to get the real results,it is necessary only to scale-up the example. The scanning spot just fills the line width, and since there are 10 units vertically (given by the number of lines), balanced horizontal definition calls for a like number of units horizontally. In this

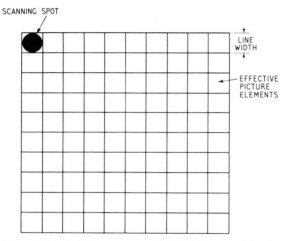

Figure 1.3. A picture composed of 100 elements. Ideally, the scanning spot is the exact size of each picture element

example, we have 100 elements in all. (Ten lines on a 50 cm picture tube would call for a very large scanning spot and, since the spot cannot define picture detail of smaller size than itself, the resulting picture would be of very poor definition.) So what decides the horizontal definition, which is established initially by the number of lines (since the

Figure 1.4. A television line of maximum bandwidth

10

horizontal definition should desirably match the vertical definition). Consider the most difficult picture or pattern to reproduce, which is alternate black and white elements (*Figure 1.4.*). The controlling signal for the scanning-spot

Figure 1.5. The video waveform for the TV line in *Figure 1.4*

brightness to display this pattern over one line has to change from maximum (corresponding to white) to minimum (black) five times. The signal waveform is represented in *Figure 1.5.* This is a squarewave. One positive-to-negative waveform as shown in *Figure 1.6,* would thus handle a black and a white

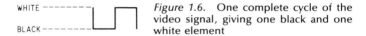

Figure 1.6. One complete cycle of the video signal, giving one black and one white element

element, so a 100-element picture would need 50 such signal cycles to deal with alternate black and white elements over the entire picture.

Bandwidth
Suppose that this 100-element picture is geared to a frame frequency of 25 Hz. The total number of black-to-white waveforms per second would then be equal to 50 times 25, giving 1 250. This represents a squarewave repetition frequency of 1 250 Hz. The vision channel would need to pass these 1 250 Hz squarewaves without distortion. In theory, the bandwidth would need to be about 10 times 1 250 Hz, or 12 250 Hz, to secure perfect transitions from black to white and from white to black, as in *Figure 1.4,* but in practice a smaller bandwidth is considered adequate, for one would rarely, if ever, wish to transmit a chequerboard pattern over the whole of the screen.

A vision bandwidth 10 times the squarewave repetition frequency is theoretically required because a squarewave is composed of a fundamental frequency sinewave plus a series

11

of sinewaves equal to the successive odd harmonics of the fundamental. When these harmonics are added to the fundamental sinewave in correct amplitude and phase relationships, a squarewave signal results, and the greater the number of harmonics added, the better the squarewave shape. If a squarewave signal is passed through a channel whose bandwidth is below that required to carry all the harmonics, distortion results and the corners of the squarewave are rounded. If the bandwidth is equal only to the squarewave's fundamental frequency, the output signal is of sinewave character. This, of course, is because all the higher-order odd harmonics are suppressed in the channel.

A 625-line picture displayed on an ordinary picture tube is not square: its width is greater than its height. The width-to-height ratio, the 'aspect ratio', is 4:3 (meaning that the width is four units and the height is three units).

To find the number of picture elements involved in such a picture, we have to multiply the product of 575 and 575 by $\frac{4}{3}$. We have already seen that $575 \times 575 = 330\,625$. Multiplying this by $\frac{4}{3}$ works out to a fraction over 440 833 picture elements in the whole picture.

What is the vision-channel bandwidth required for this number of elements? There are 25 frames or complete pictures per second, so if we divide the number of picture elements by two (because one signal caters for two adjacent picture elements) and multiply the result by 25, we obtain a figure for vision-channel bandwidth in Hertz. By taking the number of elements as 441 000, a round figure, we arrive at a bandwidth of 5 512 500 Hz, or a little over 5.5 MHz. This is representative of sinewave signal, and since we are now dealing with incredibly small elements, no larger than the diameter of the scanning spot, increasing the vision frequency to retain a squarewave characteristic would certainly not be warranted. Indeed, the authorities only transmit up to about 5.5 MHz on the 625 standard, so there is no point in arranging for a set's vision channel to have a bandwidth in excess of this.

This bandwidth has to be maintained throughout the transmission system, from the camera-tube target to the picture-tube cathode. As we shall see in subsequent chapters, the chain can be a long one, and someone once likened the preservation of a television signal to carrying a clean white shirt through a coalmine, and delivering it spotless at the exit.

The picture tube

It is only of recent years that the term 'picture tube' has been used. Previously, the display device was generally known as a *cathode-ray tube* (CRT). Indeed, the picture tube is nothing more than a large and specialised type of CRT.

Modern picture tubes have relatively flat rectangular screens with an aspect ratio of 4:3, to match the transmitted picture. The three main parts of the tube envelope are the *screen*, the *neck*, and the *flare* (cone or bulb). These are arranged in the well-known form shown in *Figure 1.7*.

The inner side of the screen is coated with so-called *phosphors*, which have the property of glowing, or *fluorescing*, when bombarded by high-velocity electrons. The neck contains the *electron gun*. This is an assembly designed to create and 'shoot' a beam of electrons at high speed on to the phosphor screen (hence the term *gun*). The flare of the tube gives room for the electron beam to be deflected without obstruction.

We have seen that external coils are fitted to the tube neck to deflect the beam vertically and horizontally. To bring the beam to a point of focus at the phosphorescent screen, an electron lens is fitted to the gun assembly. The electrons from the gun are 'pulled' through the tube towards the screen by a high positive charge on the final anode of the tube. This extra-high tension (EHT) potential accelerates the electrons to a great velocity; it is between 8 kV and 21 kV for monochrome tubes, depending on screen size.

Because electrons have a definite mass, even though they are incredibly small, their high velocity causes them to build up kinetic energy on their way from the gun to the tube. This is rather like the energy built up in, say, the head of a pickaxe when it is swung towards the ground over the shoulder. The kinetic energy of the axehead causes the point easily to penetrate the ground. Applying a similar amount of energy and endeavouring to press the point of the axehead into the ground without swinging it through the air would not work. As the axe is swung, it builds up energy, and this is liberated all at once when the point of the axe strikes the ground.

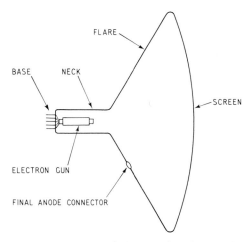

Figure 1.7. A typical picture tube, showing its main features. The inner side of the screen is coated with phosphors, which glow when bombarded by electrons from the gun

In a similar manner, the energy buildup of the beam electrons is liberated when they impinge upon the screen phosphors. The electrons cannot penetrate the screen, but instead the liberated energy is exchanged for light radiation.

This is the characteristic of the phosphors, which are a mixture of blue-emitting material (silver-activated zinc sulphide) and yellow-emitting material (silver-activated zinc-cadmium sulphite). The proportions of these are chosen to give light of the desired wavelength or *colour-temperature*.

The white required is known as 'illuminant D' and corresponds to a colour temperature of 6 500°K, the light wavelength given off by a solid when heated to this temperature. The colour is difficult to describe but is a slightly warm white. Illuminant D is of great significance in colour TV, and in Chapter 6 it will be described how three primary colours are mixed to produce this white.

It is necessary that the tube be evacuated of air for it to operate. Any residual air would give rise to *ion* production when collisions occured between gas molecules and the electron beam. Such ions are heavy and harmful, positive ions bombarding the cathode and negative ions impinging on the screen. In tube manufacture, two precautions are taken against this. A *getter* is fitted, and 'fired' once the tube is sealed to absorb residual gases and those liberated by the metal electrodes in use. Also, the technique of aluminising is used. This consists of applying a thin film of aluminium behind the phosphor screen, deposited by vaporising a pellet of pure aluminium in a vacuum. It has the twin benefits of preventing ion burn to the phosphors (a perfect vacuum cannot be achieved) and of increasing brightness by reflecting forward light which would otherwise be scattered inside the picture-tube bowl.

A tube in which there is excessive gas (air) exhibits a high beam current (since the beam is composed of ions in addition to electrons) and poor brightness and focus. Such a tube is often referred to as 'soft'.

Thermionic emission
All material consists basically of atoms, each consisting of a nucleus with one or more electrons. The electrons may be regarded as negatively charged particles, and at any temperature above absolute zero they are in orbit around the nucleus. The orbital speed of the electrons is related to the temperature of the material, rising with increasing temperature. If a suitable material is heated until it is incandescent, some electrons will acquire enough kinetic energy to break free of the material's surface and form a 'cloud' in the immediate vicinity of the glowing material. This cloud of

electrons, or *space charge*, is the source of the electron beam in the picture tube.

The electron gun

The term *electron gun* describes the entire system of electrodes which generate, focus, and accelerate the electron beam. Many types of CRT electrode system have been used in the past, starting with gas and magnetically focussed triodes. Gun systems for use in colour TV tubes have likewise undergone several stages of development. The gun system of a modern monochrome picture tube is illustrated in *Figure 1.8*. The *cathode* is in the form of a cylinder, closed at one end, and containing a *heater* – a thin spiral of tungsten in close contact with, but insulated from, the nickel cathode. A

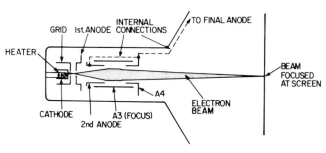

Figure 1.8. Internal arrangement of a monochrome picture tube, showing the beam path through the gun and electron lens

current is passed through the heater sufficient to raise the temperature of the cathode to a dull red glow. The outside of the cathode is coated with a mixture of barium oxide and strontium oxide, which are good emitters of electrons. The electrons thus 'boiled off' the cathode surface form the space charge between the cathode and the next electrode, the *grid*. This consists of a larger cylinder with one end closed save for a small circular aperture in the middle, and it completely surrounds the cathode.

The potential on the grid is normally held negative of that on the cathode, and when it is sufficiently negative, it repels all the electrons in the space charge back towards the cathode. This means that none can emerge through the hole in the grid; in these circumstances, the tube is said to be 'cut off', and there is no illumination on the screen. This normally occurs when the grid-cathode potential is about $-50\,\text{V}$. As the grid potential is made to rise towards that of the cathode, however, some electrons are able to escape through the grid aperture and bombard the screen. Thus the grid-cathode potential determines the intensity of screen illumination.

The electrons that emerge from the grid, having a negative charge, are immediately attracted by the high positive potential on the next electrode along, the first anode. This accelerates the electron beam and is operated at a few hundred volts with respect to the cathode. The electron beam now enters an electron lens, consisting of three anodes of cylindrical form. The first and third of these are connected together and to the final anode of the picture tube. As the beam passes through the lens, it is made to converge, and by varying the potential on the middle cylinder (focus anode), the point of convergence can be made to coincide with the phosphor screen. *Figure 1.8* shows the shape and path of the electron beam through the gun system.

The final anode is operated at 10 kV–25 kV depending on the size and type of tube. It takes the form of a conductive coating (usually of graphite and known as *aquadag*) over the inside of the CRT flare and also the layer of aluminising behind the phosphors. This high potential greatly accelerates the electrons in the beam, so that they impinge on the screen with a final velocity of many millions of miles per hour.

Beam deflection

Mounted on the neck of the picture tube are the scanning coils. This assembly, sometimes called the *yoke*, is made in a saddle-shape to fit the contour of the neck and extends a little way up the flare. The scan yoke contains one pair of scanning coils for horizontal (line) deflection, and one pair for vertical (field) deflection of the electron beam. Each pair

17

of scanning coils generates lines of magnetic force across the axis of the tube neck, as shown in *Figure 1.9*. It is a characteristic of magnetic deflection that the electron beam, in responding to a magnetic field, moves at right angles to the lines of force in that field. Thus for vertical deflection, the coils are on each side of the neck, producing horizontal lines of force; the horizontal deflection coils are fitted above and below the neck, to generate vertical lines of force.

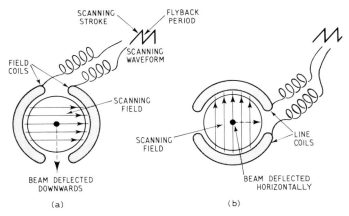

Figure 1.9. The principle of magnetic deflection. (a) Vertical deflection. (b) Horizontal deflection

The strength of the magnetic field necessary to deflect the beam over the whole screen area depends on several factors. The greater the diameter of the tube neck, the more deflection power will be required. The EHT voltage applied to the final anode is also a significant factor; as the EHT voltage is increased, the electron beam becomes 'stiffer' and thus requires more deflection power to scan through the required angle.

Since the coils are energised by a sawtooth current waveform, the beam is deflected linearly over the scanning stroke, and very swiftly returned by the retrace (flyback) stroke.

Scanning angle
The scanning angle is that angle over which the beam is deflected to cause the light spot on the screen to 'see' from one corner of the active picture area to the opposite corner. So for a given screen size, the scanning angle is increased as the length of the tube is reduced. More scanning power is required to deflect a beam of given 'stiffness' (i.e., constant EHT voltage) over, say, 90° than over 60°, so in wide-angle picture tubes, more current is required in the scanning coils, other factors being equal.

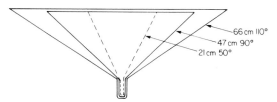

66 cm 110°
47 cm 90°
21 cm 50°

Figure 1.10. Deflection angle. For a given tube length (hence cabinet depth), the wider the deflection angle, the larger is the screen

Large and medium-size picture tubes (above 25 cm) usually have a scanning angle of 110° or 114°. For small screen sizes and portable receivers, scanning angles are usually less than this, varying down to 50°. In these applications, energy is at a premium, and the thin neck, low EHT, and small scanning angle combine to make small tubes very economical in terms of operating power.

The advantage of wide-angle picture tubes is that they can be relatively short, keeping the depth of the television cabinet to a minimum, as shown in *Figure 1.10*.

Tube construction
The large glass *envelope* of a picture tube renders it very vulnerable to physical damage, because of the great air pressure on the envelope caused by the internal vacuum. Sharp angles and abrupt changes in the cross section of the envelope are the points of greatest stress, and for this reason

19

television picture tubes need to be handled with great care. If the envelope is broken, there is a risk that atmospheric pressure will cause it to implode, and personal injury can result. Goggles and gloves are therefore recommended when dealing with picture tubes.

Precautions also have to be taken against the risk of injury to the viewer if implosion occurs. In early receivers, these took the form of a transparent barrier of one type or another. Modern picture tubes have either an integral shield or a reinforced faceplate.

For large-screen monochrome sets, 61 cm and 50 cm (24 in and 20 in) tubes are the norm. In the smaller screen sizes there is little standardisation and screens as small as 2.5 cm (1 in) have been produced, but sizes around 30 cm (12 in) are most popular. All sizes quoted are the diagonal measurement of the glass screen itself (see *Figure 1.11*).

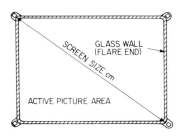

Figure 1.11. The quoted size of TV screen is the diagonal distance over the front of the glass envelope

Many current picture tubes, both colour and monochrome, are fitted with 'quick-heat' cathodes. In this form of construction, the heater element is in the form of a flat ring rather than a spiral, and the cathode is physically smaller, thus having less thermal inertia than its early counterpart. This device is capable of producing a viewable picture within five seconds of power being applied, a great improvement.

Picture centring and pincushion correction
When the picture tube is being scanned, the raster may not be centred on the CRT screen because of tolerances in the positioning of the gun within the tube neck. To correct for

this and enable the picture to be centred, a pair of ring magnets is fitted just behind the scanning-coil assembly. These are provided with tabs for manipulation and may be of metal or iron-oxide impregnated plastic, illustrated in *Figure 1.12*. The magnetic field created by these rings can be varied in amplitude and direction by rotating them relative to the tube and to each other, so that the electron beam, before entering the deflection field, can be 'bent' to centre a test-card display on the screen. Any tilt on the picture can be corrected by rotating the complete scanning-coil assembly.

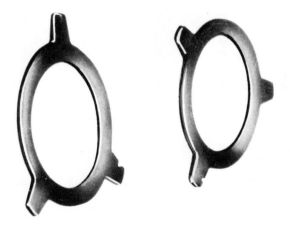

Figure 1.12. Picture-shift magnets, which are used to centre the picture on the screen, are usually rings of flexible ferro-plastic material

In wide-angle picture tubes, the relatively flat screen and large deflection angle cause the raster to be bowed-in at the edges, as shown in *Figure 1.13*. This is known as *pincushion distortion*. It is more pronounced at the sides than at the top and bottom because of the 4:3 aspect ratio and hence wider scanning angle for horizontal deflection. Correction is made by means of bar magnets attached by flexible arms to the

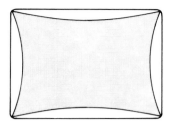

Figure 1.13. Pincushion distortion of a raster

front (flare end) of the scanning-coil assembly. Two or four may be fitted, and the resultant magnetic field is such that it pre-distorts the raster edge to compensate for the pincushion effect. The magnets are adjusted for straight edges on a test card or crosshatch display.

Aquadag capacitance
Connection to the final anode of the tube is by a cavity connector on the flare of the tube. This gives plenty of clearance for the high voltage present, and its leads through the glass to the internal aquadag coating already described.

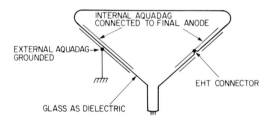

Figure 1.14. Conductive coatings (aquadag) on each side of the glass flare form a convenient EHT reservoir capacitor

The outer surface of the flare is also coated with conductive aquadag, and these two layers form a capacitor, with the glass of the tube flare as dielectric. In large tubes, the capacitance thus formed can be as great as 2 000 pF and as such is a useful reservoir for the EHT voltage when the external aquadag is 'earthed' to the negative line or metal chassis of the set: the idea is depicted in *Figure 1.14*. In many sets, this makes it unnecessary to employ a separate, high-voltage capacitor for this purpose, the exceptions being some small-screen receivers with a correspondingly small area of aquadag.

Developments in picture tubes

A great deal of research is going on, and has been for many years, in the field of flat-screen displays for television. This principle eliminates thermionic emission and magnetic scanning, individual picture elements each being 'addressed' by the drive system. This is very close in principle to the bulb matrix system with which television was introduced at the beginning of this chapter, the functions of scanning and intensity modulation being integrated. This nut is a hard one to crack, the main problems being illumination intensity and the sheer number of elements necessary for a full-definition picture – as we have seen, the 625-line system requires almost 450 000 individual picture elements. Small-screen prototypes have been demonstrated, with much reduced definition, in photo-emissive (LED) and reflective (LCD) form. The nature of these displays is similar to those we are familiar with in watches, clocks, and calculators. There is no doubt that it will be some years before these experiments come to fruition, but they surely will one day, and then, except for special purposes, thermionic emission will be obsolete.

Another approach to flat-screen technology is the display developed and marketed by the Sinclair company. This is a modification of the traditional system, still using a heated cathode in an evacuated glass envelope, but having a much more practical shape. *Figure 1.15* shows the idea. The electron gun is mounted parallel to the viewing window and phosphor screen. Vertical and horizontal deflection is carried out electrostatically while the beam is moving along the major axis of the tube (box?). The electron beam strikes the phosphor screen from the same direction as the screen is viewed, having been deflected through 90° when it came into the area of the screen. This 90° deflection is again achieved electrostatically by a positive charge on the phosphor screen itself and a negative charge on a transparent conductive coating on the inside of the viewing window. The system is fraught with inherent raster distortion, and this is corrected electrically and optically. A form of electron lens, called a 'collimator', is introduced into the electron gun. This, in conjunction with the application of correcting signals to the electrostatic deflection plates, removes *trapezoidal* (keystone-shape) distortion from the raster. Further, optical, correction is given by the effect of the Fresnel lens of the

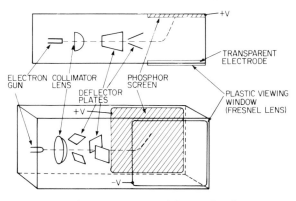

Figure 1.15. The arrangement of the Sinclair flat picture tube. The electron gun is parallel to the plane of the phosphor screen and viewing window

viewing window. Apart from its physical advantages, this tube has a bright display and is very efficient in energy input. Also, manufacturing costs are low because there are no wound accessories. The tube has been made in 75 cm (3 in) size for use in small portable TV sets.

Summary

This chapter has explained how television involves breaking up the image to be transmitted into picture elements, and

24

scanning them in orderly sequence for transmission to the receiver in the form of a train of pulses, finally to be reassembled on the screen of the picture tube.

Let us recapitulate on the controls required by the picture tube. We need a heater current to cause the cathode to emit electrons. To form the high velocity beam, we need an EHT supply for the final anode and a smaller voltage for the first anode. Also required is a focussing potential of 0 to 300 v or thereabouts. We need a negative grid voltage, regulated by the brightness control, to adjust the scanning spot and so the brightness of the picture. And the vision signal (usually called *video* signal) is applied between the grid and cathode of the tube to change the brightness of the scanning spot in sympathy with the picture elements encoded on the transmission. On the tube neck, we need coils to generate magnetic fields for spot deflection, and possibly further permanent magnets to centre the picture and correct distortion of the shape of the raster.

2

Transmission

A television camera possesses a lens or optical system just like that in an ordinary film camera. The optical system focuses light rays from the scene so that its image appears inverted on the sensitised screen of the camera tube. The camera as a whole contains not only the camera tube but electronic equipment to control it. Thus the camera tube is rather like the picture tube in a television set, which can work only under the control of its associated electronic ciruits.

Like cathode-ray display tubes, the camera tube has evolved through many forms. For many years, *photoemissive* tubes were used for broadcasting purposes. In this type of tube, the sensitised screen develops an electric potential when light falls upon it, and this potential is picked up and amplified within the tube. The final form of this tube (image orthicon) is physically large and electrically complex, and it has now been superseded by tubes of the *photoconductive* type, which are simpler in operation and much smaller.

Photoconductive camera tubes

The original photoconductive camera tube was the vidicon. In its early forms its performance fell short of other types of tube, notably in the areas of sensitivity and 'lag', the latter giving rise to a 'comet tail' effect on moving objects. Later versions, such as the plumbicon, overcame these problems and are almost exclusively used by broadcasters today.

Chapter 1 made it clear that a television scene cannot be transmitted as a whole. It has first to be broken down into elements by a scanning process to allow the elements to be transmitted sequentially. It is the function of the television camera to scan the focused image and to translate the picture elements into a series of electrical impulses.

Let us see how this takes place in a camera system using a photoconductive tube. The basic elements of the vidicon are shown in *Figure 2.1*. Its action is the reciprocal of that of the picture tube. That is, it is designed to produce video signals from a picture, while the job of the picture tube is to produce pictures from video signals. In audio the microphone and loudspeaker are similarly related.

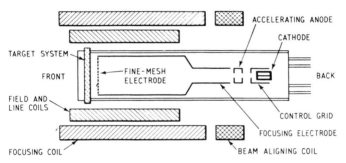

Figure 2.1. The main elements of a photoconductive camera-tube system

The sensitised screen in a typical vidicon, however, is only about 25 mm (1 in), while the screen of a picture tube may be as large as 67 cm. This means that the image of the scene focused on the vidicon is scaled-up on the picture tube.

The vidicon has three main elements. These are an electron gun, the beam scanning system, and the 'target area', upon which the electron beam impinges and upon which the optical system focuses the scene being televised. It will be recalled that in the picture tube, the electron beam strikes the fluorescent screen to produce the scanning spot. The beam in the camera tube, however, does not result in

27

illumination but instead 'reads out' information concerning the brightness of the image upon the target area at any point upon which the beam impinges at any instant. Thus, as the beam scans the target area, it causes the production of impulses which carry information as to the monochrome shading of the image. This is effectively the translation of the picture into picture elements of an electrical nature.

The electron gun in the vidicon works in a similar manner to that in the picture tube, described in Chapter 1, the beam electrons being brought to a high velocity by the first anode potential of about 300 v. For optimum resolution, the beam is focused by an electromagnet so that it impinges in a sharp point on the target area, and it is also deflected vertically and horizontally by field- and line-scanning coils, as in the picture tube. In Great Britain, the repetition frequencies are (field) 50 Hz and (line) 15 625 Hz on the 625 standard, matching those used for scanning in the receiver. *Figure 2.1* shows the basic gun; the focusing coil, the line and field scanning coils, and a beam alignment coil is used to compensate for minor tolerances in the construction of the tube.

It will be seen that there is also a focusing electrode in the gun assembly. This is used in conjunction with the magnetic field for beam focusing and facilitates beam adjustment at the camera.

In effect, the principles of scanning are virtually the same as in the picture tube. There are minor differences in detail, as would be expected. One is that at the end of the cylindrical focusing electrode is an electrode made of fine wire mesh, the purpose of which is to retard the electrons just before they impinge on the target area and to form a barrier to prevent ions from reaching the target.

The scanning coils are energised by sawtooth current waveforms of an amplitude sufficient to fully deflect the beam and at repetition frequencies to match the transmission standard. Thus, the focused end of the beam scans the sensitised area in exactly the same way as the scanning spot in the picture tube scans the fluorescent screen. One requirement of the television system is that the beam in the camera tube must deflect both vertically and horizontally in exact sympathy with the beam deflection in all the receivers using the transmission. This is called *synchronisation*, and is dealt with in detail later.

The target

The optical end of the camera tube is illustrated in *Figure 2.2*, which shows that the scene to be televised is focused by a lens system upon the front of the target area or conductive film.

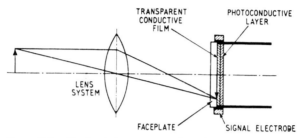

Figure 2.2. The basic lens system of a vidicon camera tube

This scene image is totally scanned by the focused electron beam, an action which is fundamental to any camera tube. How does the tube translate the image into electrical impulses? The answer lies in the nature of the photoconductive material comprising the target area and is based on the fact that the material has many of the features of a light-dependent resistor (LDR). That is, its electrical resistance is very high when the intensity of light falling upon it is low (giving the so-called *dark resistance*) but decreases fairly swiftly as the light intensity increases. In the first vidicons, the target coating was selenium, but its performance was poor, and antimony trisulphide became the favoured material. This had a better panchromatic response, but eventually it gave way to lead oxide as a target material. This is capable of much better 'lag' performance and, except for surveillance and similar non-critical applications, has largely superseded the other materials.

As a result of the scanning action of the electron beam, the photoconductive film or layer can be considered as being broken down into a large number of very small resistive elements, across each of which is a capacitance created by the photoconductive material itself as one plate and the transparent conductive film (see *Figure 2.3*) as the other plate.

So the target area can be considered to be a very large number of small resistors, each in parallel with a capacitor. One side of the combination is the common target electrode, while the other side of each resistor/capacitor (RC) pair is effectively open-circuit until the electron beam comes into play, as we shall see.

Since the electron beam scans the complete target area, each RC pair is connected in turn to the beam for a very brief moment; because the beam is composed of electrons, it is the equivalent of an electric conductor carrying current. But, while an ordinary conductor (a copper wire, for instance) is somewhat inflexible, a conductor consisting entirely of electrons, such as the beam, is considerably more flexible and is easily deflected without the mechanical difficulties associated with any other kind of conductor. *Figure 2.3* shows a circuit equivalent to that of the photoconductive film with RC pairs and the electron beam. This shows that the 'common' side of the RC pairs is connected to a positive potential through a load resistor.

When the beam falls upon the RC pairs, each in turn, as it scans the photoconductive material, the capacitive elements acquire an electric charge. The charging circuit is from the positive side of the load resistor on the target (or *signal electrode*, as it is sometimes called) through the load resistor, across the capacitance, through the electron beam and gun, and back to the negative side of the signal electrode potential.

Since the capacitance of each RC pair is shunted by a resistance, the charge acquired via the beam will start to leak away as soon as the beam leaves the pair. As soon as the beam leaves a particular pair, the charge picked up by the capacitance will immediately start to leak away through the

30

parallel resistance, and by the time the beam comes round to this pair again on the next scan, the capacitance will have lost a certain value of charge, and a corresponding amount of current will have to flow through the load resistor fully to restore the charge.

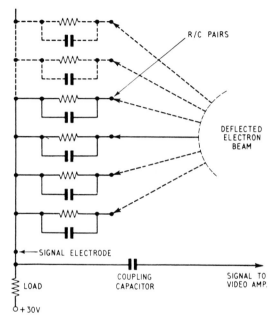

Figure 2.3. The equivalent circuit of the scanned points of the photoconductive target area shown as resistor/capacitor pairs

Suppose that the camera's lens cap is in position, meaning that there is no light focused on to the target. The photoconductive material thus has a high *dark resistance*, so virtually no charge will be lost from the capacitance of each RC pair, and in consequence no current will flow through the load resistor as the target is scanned, since there is no charge loss to boost.

31

On the other hand, if the camera is looking at a patch of bright light, a bright image will be focused on to the target and the photoconductive material will have a much lower resistance at each RC pair, so a substantial charge current will flow through the load resistor as the beam scans each RC pair.

The video signal at the target

The signal representing the various shades between black and peak white of an image focused upon the target thus consists of the pulses of current through the load resistor as the beam scans the target area. These current pulses produce voltage pulses of equivalent character across the load resistor connected to the signal electrode, and it is this signal voltage that represents the vision or *video signal* delivered by the camera. *Figure 2.3* shows that the signal is fed from the signal electrode to a video amplifier via a coupling capacitor which isolates the DC in the load.

So the image of the scene televised is scanned by an electron beam and broken down into elements corresponding to electric pulses whose magnitude is governed by the relative brightness of each element. Black gives very little or no video signal, while peak white gives maximum output. A shade between black and white produces a voltage whose level falls somewhere between zero and peak-white levels. This represents a grey in monochrome television.

It will be remembered that the image of the scene is scanned in successive lines to make up one field, and that there are two interlaced fields to one frame or complete picture.

If the scene is a series of alternate black and white vertical bands, one line of video signal might appear like the waveform in *Figure 2.4 (a)*, while the signal waveform resulting from three successive lines of an average picture image could appear like that in *Figure 2.4 (b)*. In other words, the *luminance* of the image focused upon the sensitised area of the camera tube is translated line by line into lines of video signal.

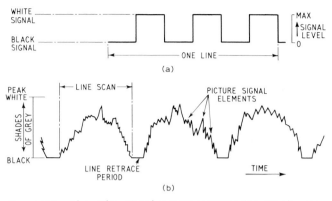

Figure 2.4. The video signal output representing (a) alternate black and white vertical bars and (b) three lines of a typical image

Lead-oxide tubes are capable of providing first-class pictures and ultra-sensitivity in modern camera systems, some producing pictures down to 50 lx (50 lumens per square metre), with normal working at 500 lx.

Flying-spot scanner

There is another way of generating picture signals which does not involve a camera tube at all. This is the *flying-spot scanner*, which can televise transparences and cine-film. A small (75 mm–100 mm, 3 in–4 in) cathode-ray tube is set up and provided with EHT, scanning fields, and so on, just as in a picture tube. This produces a small, rectangular, and intensely bright raster. The beam in the tube is not modulated, so that the raster on the faceplate is quite blank. For this application, it is important that the *afterglow* (decay time) of the tube is very short, and a short-persistence purplish-blue phosphor is utilised. The raster thus produced is focused by an optical system on the transparency to be televised, so that it exactly fits the image size. The setup is shown in *Figure 2.5*.

If we imagine ourselves on the outside of the transparency, or slide, we would see the raster 'through' the slide, and if only

33

our eyes were fast enough, we would see but a single spot of light scanning the slide. When the light spot passed over an opaque part of the slide, its intensity would fall, while clear areas of the slide would allow the light beam to pass unimpeded. All we need now is a photocell to view the scene, and because the light coming from the slide is modulated with picture information, the electrical output from the photocell would be a video waveform, an electrical analogy of the slide image, just like that from the camera

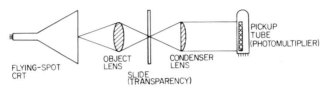

Figure 2.5. The optical setup for slide scanning in a flying-spot scanner

tube. A condenser lens is interposed between the slide and the photocell. This prevents scattered light from being wasted, by concentrating it on the photosensitive surface of the cell. Care is taken that the image is not actually in focus at the cell's surface, however, because any imperfections here would then be reproduced on the picture.

We can see now why the flying-spot scan tube must have a short afterglow. If any decaying illumination is present after the spot has moved on, this will be seen by the photocell and cause loss of definition – and lag where moving ciné-film is being reproduced. For the same reason, the photocell must have a fast response, and here, in contrast to camera-tube practice, photoconductive cells (selenium, etc) are far inferior to photoemissive devices, which are also able to offer internal amplification of the signal.

Photomultiplier

The light output from the slide, then, is concentrated on the cathode of a photomultiplier. This is a 'cold' tube, having no heater and consisting of a series of electrodes mounted in a

34

cylindrical glass envelope. The construction is shown in *Figure 2.5*. There are typically nine electrodes, or *dynodes*, in the photomultiplier tube and, starting with the cathode (the one which 'looks out') at a low potential, each is held at a higher voltage than its predessessor. This is achieved with a chain of resistors fed from a high-voltage source.

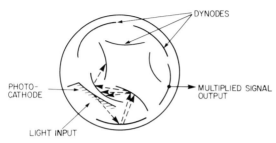

Figure 2.6. Internal electrode structure of a photo-multiplier

When light strikes the photosensitive cathode, it emits electrons, and these collide with the surface of the first dynode. Because this is some tens of volts higher in potential than the cathode, the bombarding electrons 'knock off' electrons from the dynode, a phenomenon known as *secondary emission*. These emitted electrons are greater in number than those which triggered them off, and they collide with the surface of the second dynode, where secondary emission also takes place. This process is repeated, avalanche style, through the dynode chain, so that the final dynode picks up a large signal, forming the video output. The device is fast enough to faithfully reproduce every detail of the image on the slide.

Flying-spot scan systems are capable of very good repro-duction and lend themselves to test-card and cine-film repro-duction, being called *telecine* systems in the latter case. They are particularly practical in colour applications as they avoid the problem of *registering* the three colour images (more of this in Chapter 6). A great deal of television output comes from flying-spot scanners of one kind or another, though for test-cards, captions, and even such images as moving clocks, all-electronic generators are now becoming common.

The video signal

So far we have seen that the camera, whatever form it takes, produces lines of video signal as the beam scans the image to be televised. If these signals were used to modulate the grid or cathode of a picture tube, provided the scans were synchronised, the scanning spot on the fluorescent screen would alter in brightness in exact accordance with the brightness in exact accordance with the brightness or luminance value of the picture elements scanned by the camera, thus building up an exact reproduction of the image at the studio.

Synchronising the scans
Synchronising ('sync' for short) means ensuring that the scanning beam at the picture tube is in exact step with that in the camera tube at the studio. They start together at the top left-hand corner, and at any given moment they are at the same position on their respective screens. Information has to be transmitted along with the picture waveform to synchronise the scans and to keep them in step.

This is achieved by adding sync pulses to each line of video signal to keep the line scans in step, and to each field of video signal to keep the vertical scans in step. In *Figure 2.4 (b)* can be seen gaps betwen each complete line of video signal, and it is during these periods that the line flyback occurs. It will be seen that the signal amplitude is at black level in these intervals, which means (theoretically) that the picture-tube beam current is cut off and the scanning spot is extinguished. So it is possible to include line-sync pulses in these intervals without there being any indication of these on the picture, provided the pulses do not cause a rise in video signal amplitude above the black level.

To avoid the pulses affecting the picture (that is, to make sure they do not cause a scanning spot to appear on the picture tube), they are arranged to *fall* in amplitude *below* the black level of the video signal itself. In television parlance, below-picture black level is called *blacker-than-black*: three line-sync pulses, each following a line of video signal, are shown on the waveform in *Figure 2.7*.

It is shown in Chapters 4 and 5 how these (and the field-sync pulses – see later in this chapter) are separated from the true picture signal at the receiver, how the picture signal controls the brightness of the scanning spot, and how the sync pulses 'lock' the vertical and horizontal scanning timebases to the accurately controlled timebases at the transmitter. This last process is called *synchronisation*.

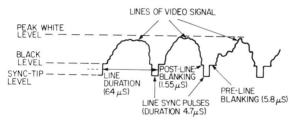

Figure 2.7. Three lines of video signal with sync pulses

The waveform in *Figure 2.7* corresponds to the 625-line standard, and it will be seen that the duration of one complete line of signal (including a line-sync pulse) is 64 μs (μs = microsecond = 0.000001 sec) and that the duration of a line-sync pulse is 4.7 μs. Another important point is that the start and finish of each line of video signal does not fall immediately into the region of blacker-than-black to form one side of the line sync pulses. Instead, each line starts and finishes for a very short period at black level. These periods correspond to the post-line and the pre-line blanking (sometimes called the front and rear 'porches' of the line-sync pulses). Their purpose is to give the sync and timebase circuits time to respond to the change in signal amplitude from video to sync, especially if a line starts or finishes at peak white. Without these intervals, the signal would in the latter case have to change almost instantaneously from peak white to blacker-than-black – an impossibility. The maximum change with the intervals present on the signal waveform, therefore, is from peak white to black, and the intervals themselves give the circuits sufficient time to respond to such a sudden and sharp change in signal amplitude. Intervals respectively of 1.55 and 5.8 μs have been adopted. (We shall see when we look at colour television in Chapter 6 that the post-line interval carries colour-sync signal waveforms.) The waveform photograph in *Figure 2.8* shows the appearance of almost three lines of video signal at the receiver.

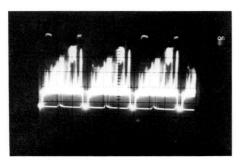

Figure 2.8. Waveform photograph, showing just over three lines of video signal at the receiver

The line-sync pulses serve to instigate the retrace action of the sawtooth current waveform in the receiver's line-scanning coils, and in this way the line scans at the receivers commence at exactly the same instant as those at the television camera. The camera-tube and picture-tube electron beams are thus locked together horizontally.

Field-sync pulses
The same high degree of synchronisation must also be achieved with the field scans, but this is a little more difficult

38

because the complete picture or frame consists of two interlaced fields, as explained in Chapter 1. This means that the field sync must identify the two fields of a frame as well as holding the electron beams vertically in step.

We have seen that each field consists of 312½ lines and each field has its own sync signal. Actually, instead of there being just a single pulse, as between lines, there is a train of sync pulses between fields. The idea of these is shown in *Figure 2.9*: at *(a)*, the signal at the end of the 'even' fields, and at *(b)*, at the end of 'odd' fields.

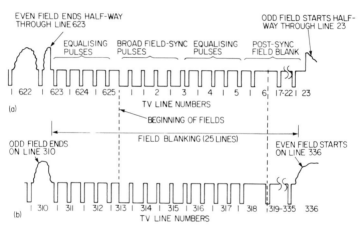

Figure 2.9. Details of the train of sync pulses between fields. (a) At the end of even fields. (b) At the end of odd fields

The diagrams show that the field-sync signals proper consist of a series of pulses, some of which are wider than the line-sync pulses, having a duration of about 28 μs. This pulse train immediately follows the last line of a television field and is interpreted in the receiver as a 'trigger' for the field flyback. The field-pulse train is followed by a series of lines carrying no picture information. This interval (about 22 lines) allows time for the field timebase to settle down before the new field is traced out.

In Chapter 1 it was explained that not all the 625 lines of the television raster are active lines, and the 'lost' ones, containing the field sync pulse train and *post-sync field blanking*, are thus responsible for synchronising the field scan. In fact, some of the spare lines, during the field blanking interval, are utilised by the broadcasters for test and data signals – more of this is Chapter 9.

A careful study of the diagrams in *Figure 2.9.* should remove the problems that many beginners at television have in understanding interlaced frames. At the left-hand side of *Figure 2.9(a)*, completion of one television frame is approaching. The last active line, no 623, terminates half-way through, and the field-sync pulse train instigates the field flyback at the receiver. The flyback cannot be achieved instantaneously, and it takes a finite time, corresponding to several TV lines, for the spot to fly back to the top left-hand corner of the screen. Let us assume that it has arrived by line 12 of the new field. The next 11 lines are traced out at the top of the screen, then the picture information starts half-way through line 23 and runs until the end of line 310, when a further series of field-sync pulses is transmitted. This brings us to *Figure 2.9 (b)*, in which it is seen that the picture signal ends on a full line (not a half one) – that is, line 310. After the sync pulse, there is another series of blanked-out lines (post-field sync blanking), and the picture signal restarts at the beginning of line 336.

Thus it is the actual timing of the field-sync pulses on odd and even lines that is responsible for the accurate interlacing of the lines of one field in the gaps between the lines of the partnering field. Because in effect the field-sync pulse arrives half a line early on alternate fields, every other field will be traced out half a line early; this will fit the lines on that field exactly between the lines of the preceding and succeeding fields as in *Figure 1.2*. It can be seen that any slight mistiming here can destroy the interlace and cause pairing or overlay of lines of two fields, thereby destroying half the vertical definition of the picture. The synchronising pulses, then, require to be generated with great accuracy and carefully preserved in the receiver.

40

An interesting point about the regularly spaced 4.7 μs intervals throughout the field-sync period is that they serve to keep the receiver's line timebase in synchronism during the field-sync periods. At the receiver, these 4.7 μs intervals are differentiated by the sync circuits, and are interpreted by the line timebase as line-sync pulses. This is explained in Chapter 5.

Pulse generation

The broadcaster uses elaborate equipment and circuits for the generation of sync and timing pulses in the studio. These waveforms are tied to the reference signals used in colour broadcasting and are derived from a single source to which all studios and transmitters are linked. Careful quality control is maintained at studios and transmitters to ensure there is no degradation of the reference signals and pulse timings.

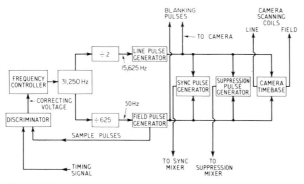

Figure 2.10. Block diagram of the pulse generator of a transmitter

A simplified pulse generator is shown in *Figure 2.10*. A crystal master oscillator running at twice line frequency (31250 Hz) regulates the accurate timing of the whole system. The line repetition frequency is simply derived by dividing the oscillator output by two, while to obtain the 50 Hz field repetition frequency a divide-by-625 circuit, known as a *counter*, is used. The line and field pulses so obtained drive their respective generators, and further circuitry provides correctly timed and shaped pulses for sync and blanking.

It will be seen that the master oscillator is under frequency control from a controller receiving a correcting voltage from a discriminator. This discriminator receives two signals, one from a timing source and one from the field-pulse generator. Provided these two signals are in step with one another, no correcting voltage is generated by the discriminator, but should the signals wander one from the other in frequency or phase an *error voltage* is produced and applied to the frequency controller. This 'pulls' the master oscillator back into correct frequency.

The line and field generators also produce the sawtooth currents for deflecting the electron beam in the camera tube, and further pulses are fed from the network to blank or cut off the camera-tube electron beam during the line- and field-retrace periods. These pulses are fed to the grid or cathode of the tube gun, negative-going in the former case and positive-going in the latter.

Composite video signal

The composite video waveform is developed by combining the video signal from the camera and the pulses from the pulse generator. The block diagram in *Figure 2.11* shows basically how this is achieved. The video signals from the

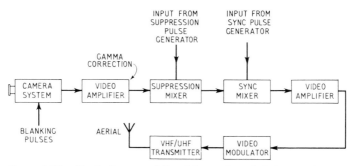

Figure 2.11. Block diagram of the video and modulating sections

42

camera system are first *gamma-corrected* to provide a pan-chromatic characteristic and amplified. They are then passed through the suppression mixer which deletes the picture signal during the sync periods, as we have seen, and thence through the sync mixer, which introduces the sync pulses themselves, while leaving the front and back porches of the line-sync pulses (i.e., giving pre- and post-sync line suppression). Finally, the composite signal is further amplified to a level suitable for driving the video modulator at the transmitter.

It should be understood that composite waveforms such as those shown in *Figures 2.7.* and *2.9* represent the *modulation signal*, as is present at the video modulator at the transmitter, for instance. Similar signals, but carrying a greater level of distortion, are developed across the load of the vision detector at the receiver. These signals (see Chapter 4) are amplified by the set's video amplifier and passed to the picture tube and sync separator.

Video modulation

At the transmitter, the composite video signal is modulated on to a radio-frequency carrier wave for transmission from the aerial system. The principles of television transmission are similar to those of radio transmission, but there are a number of differences in detail.

We have seen that to retain the transient and pulse nature of a video signal, bandwidths up to about 5.5 MHz are required on the 625 standard. Substantially smaller bandwidths will reduce the horizontal resolution of the picture. Composite video signals can be looked upon as modulation signals of sinewave components extending up to about 5.5 MHz.

If these are modulated in the same way as audio signals, sideband signals would be developed which, at maximum resolution of the system, would extend 5.5 MHz above the carrier frequency (the upper sideband) and 5.5 MHz below (the lower sideband). Thus a portion of the radio-wave spectrum twice the maximum modulation frequency would be required to accommodate just one vision channel. Indeed, in 1936, at the commencement of television transmission using the 405-line system, this technique was used, and it was retained on the London Channel 1 transmitter for some time afterwards. However, this was so wasteful of radio space that it was impossible to carry all the projected television channels in the spectrums made available for them.

The problem was solved by transmitting only one sideband in full, with the other sideband almost completely suppressed. This is called *single-sideband modulation*. Sometimes the term *vestigial sideband* is used, indicating the presence of a vestige of modulation on the suppressed sideband, since it is impossible completely to eliminate one sideband.

In the British 625 television standard, the upper sideband is transmitted, and for this reason *upper-sideband* characteristics are often referred to. *Figure 2.12* shows respectively at *(a)*

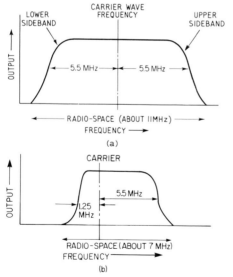

Figure 2.12. Transmission characteristics. (a) Double sideband. (b) Upper sideband, which conveys the same information

44

and *(b)* double- and single-sideband transmission characteristics. These diagrams clearly show the saving of radio space by the lower sideband technique, and it should be understood that in spite of the almost total elimination of the lower sideband, the modulation information is in no way lost or destroyed, since all the information necessary for the receiver is carried in one sideband of the carrier wave, the other sideband merely duplicating this information.

Another factor revealed by these diagrams is that ordinary medium frequencies cannot be employed for vision transmission, since over the whole of the medium and long-wave bands there would not be sufficient space to accommodate one channel. For this reason, television transmissions occupy

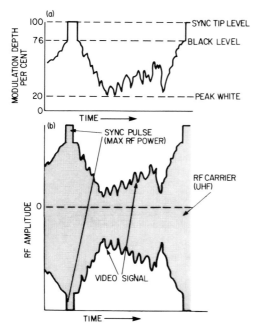

Figure 2.13. Amplitude modulation of the RF carrier. (a) A typical video signal. (b) The resulting RF envelope pattern

the *ultra*-high-frequency (UHF) bands, where there is plenty of radio space. These bands and the channels in them are detailed towards the end of this chapter.

On the British 625 standard, the video signal is modulated on to the carrier in such a way that sync tip level represents 100 per cent modulation, black level 76 per cent, and peak white about 20 per cent modulation. This is termed *negative modulation*; it is shown over one line of signal in *Figure 2.13 (a)*.

Because the vision signal influences the amplitude of the carrier wave upon which it is modulated, the signal is said to be amplitude modulated (AM). *Figure 2.13 (b)* conveys the idea. It will be recalled that AM is also used on radio transmissions occupying the medium-frequency and short-wave broadcast bands.

The sound signal

Television needs both vision and sound. The sound transmissions occupy a carrier wave which is adjacent to, but completely independent of, the vision carrier wave and is generated and modulated by its own transmitter.

The sound-transmission system, unlike that for vision, is frequency modulation (FM). This is the same as that used for VHF sound transmissions in VHF Band II. As we have seen, the upper sideband of the vision carrier signal extends to about 5.5 MHz. To keep the sound transmitter frequency clear of this, it is positioned 6 MHZ above the vision carrier, so on channel 50, for instance, the vision carrier is at 703.25 MHz, with the sound at 709.25 MHz. Because the sound channel only embraces audio modulating frequencies up to about 16 KHz, its sidebands are neglible compared with those of the vision signal. The sound transmitter does not need to be as powerful as the vision transmitter, and the ratio of vision to sound carrier powers is 5:1.

In low-power transmitters and relay stations, the sound and vision signals are both handled by the final RF power

output stage of the station. For high-power and main transmitters, however, two quite independent power amlifiers are
used, and their outputs are combined by special filters
before application to the common aerial system.

The transmitting aerial

The job of the aerial is to launch the sound and vision carrier
waves into space. The power-handling capabilities of transmitting aerials may be great, but because of the very short
wavelengths at UHF, physical size is not a problem. Very
sophisticated aerial arrays are often needed, tailored to give a
field-strength pattern suited to the local geography.

TV transmitting aerials consist of a series of dipoles or
slots, tuned to the frequencies in use and mounted atop a tall
slender tower. UHF transmitters are *co-sited*, and in many
places one aerial array radiates all the carriers for sound and
vision of all four transmitters within a local *group*.

Plane of polarisation
Radio waves are composed of an electric and a magnetic
component (hence the name *electromagnetic waves*). These
two components are always at right-angles to each other; the
direction of travel of the wave is at right-angles to both
components.

Since a radio wave is a product of oscillatory energy from
the television transmitters, the polarity of the two components is continuously alternating in sinewave form. It is the
continuously alternating nature of the magnetic and electric
components of the wave that sustains it in space.

The distance taken by one complete cycle of wave is called
the *wavelength*. This is equal to the wave velocity divided by
its frequency. The velocity of electromagnetic waves is
almost exactly equal to 300 000 000 m per second. Thus,
given either the wavelength or frequency of a radio wave, the
other can be found without trouble.

A radio wave is said to be 'polarised', and its plane of
polarisation corresponds to the direction of the lines of the

electric component. Thus, a wave generated with the lines of the electric field running vertically is said to be 'vertically polarised'. Such a wave needs a vertical receiving dipole aerial in order to abstract its energy fully, while a horizontal dipole is required to receive a wave of horizontal polarisation (whose electric field is running horizontally). Both vertical and horizontal polarisations are used in television to enhance discrimination between different transmissions working in a common or shared channel. Generally, high-power main stations generate horizontally polarised waves, while relay stations are vertically polarised.

Bands and channels

A 625-line television channel is 8 MHz wide, as illustrated in *Figure 2.14*. This is drawn for channel 52 and shows the vision carrier at 719.25 MHz with its main sideband above and

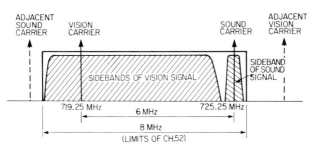

Figure 2.14. Spectrum of transmitted TV signal, drawn for channel 52

vestigial sideband below. The sound carrier is 6 MHz away, and the fact that the sound and vision signals do not quite use up the full 8 MHz available means that a small space, or *guard band*, is left between adjacent channels.

There are four active television bands for broadcast use, two in the VHF spectrum and two in the UHF spectrum. The VHF ones are called bands I and III, and the UHF ones bands IV and V. There is also band II in VHF, but this is used for FM sound broadcasting only.

The television transmitters on the VHF bands I and III and the obsolescent 405-line transmissions they carry are being phased out, and all will be gone in 1986. After this date, the use to which band I may be put is in question, but it seems very likely that the five available channels in band III will be used for 625-line television transmissions, possibly on a local basis. The handicap here is that band III receiving aerials are bulky compared with UHF types, and these will have to be installed alongside the existing band IV and V aerials to receive the new transmissions.

Band IV is divided into 14 television channels (21 to 34) and band V into 30 (39 to 68). Channels 35 to 38 are not used for

Table 2.1. UHF channels and carrier frequencies

UHF band IV			UHF band V		
Channel	Sound (MHz)	Vision (MHz)	Channel	Sound (MHz)	Vision (MHz)
21	477.25	471.25	47	685.25	679.25
22	485.25	479.25	48	693.25	687.25
23	493.25	487.25	49	701.25	695.25
24	501.25	495.25	50	709.25	703.25
25	509.25	503.25	51	717.25	711.25
26	517.25	511.25	52	725.25	719.25
27	525.25	519.25	53	733.25	727.25
28	533.25	527.25	54	741.25	735.25
29	541.25	535.25	55	749.25	743.25
30	549.25	543.25	56	757.25	751.25
31	557.25	551.25	57	765.25	759.25
32	565.25	559.25	58	773.25	767.25
33	573.25	567.25	59	781.25	775.25
34	581.25	575.25	60	789.25	783.25
39	621.25	615.25	61	797.25	791.25
40	629.25	623.25	62	805.25	799.25
41	637.25	631.25	63	813.25	807.25
42	645.25	639.25	64	821.25	815.25
43	653.25	647.25	65	829.25	823.25
44	661.25	655.25	66	837.25	831.25
45	669.25	663.25	67	845.25	839.25
46	677.25	671.25	68	853.25	847.25

Table 2.2. Band IV/V (UHF) stations and channels

Station	Channels					
	BBC1	BBC2	ITV	Ch4	Aerial group	Polarisation
Angus	57	63	60	53	C/D	H
Beacon Hill (S. Devon)	57	63	60	53	C/D	H
Belmont	22	28	25	32	A	H
Bilsdale, West Moor	33	26	29	23	A	H
Whitby†	55	62	59	65	C/D	V
Black Hill	40	46	43	50	B	H
Blaenplwyf	21	27	24	31	A	H
Brougher Mountain	22	28	25	32	A	H
Belcoo†	51	44	41	47	B	V
Caldbeck	30	34	28	32	A	H
Whitehaven†	40	46	43	50	B	V
Caradon Hill	22	28	25	32	A	H
Carmel (Carmarthenshire)	57	63	60	53	C/D	H
Chatton	39	45	49	42	B	H
Craigkelly	21	27	24	31	A	H
Crystal Palace	26	33	23	30	A	H
Guildford†	40	46	43	50	B	V
Hemel Hempstead†	51	44	41	47	B	V
Hertford†	58	64	61	54	C/D	V
High Wycombe†	55	62	59	65	C/D	V
Reigate†	57	63	60	53	C/D	V
Tunbridge Wells†	51	44	41	47	B	V
Darvel (Ayrshire)	33	26	23	29	A	H
Divis	31	27	24	21	A	H
Kilkeel†	39	45	49	42	B	V
Killowen Mountain†	21	27	24	31	A	V
Larne†	39	45	49	42	B	V
Dover	50	56	66	53	C/D	H
Durris	22	28	25	32	A	H
Eitshal	33	26	23	29	A	H
Clettravel†	51	44	41	47	B	V
Emley Moor	44	51	47	41	B	H
Chesterfield†	33	26	23	29	A	V
Halifax†	21	27	24	31	A	V
Keighley†	58	64	61	54	C/D	V
Sheffield†	21	27	24	31	A	V
Wharfedale†	22	28	25	32	A	V
Hannington	39	45	42	66	E	H
Heathfield	49	52	64	67	C/D	H

Table 2.2. Band IV/V (UHF) stations and channels (*continued*)

Station	*Channels*					
	BBC1	*BBC2*	*ITV*	*Ch4*	*Aerial group*	*Polari- sation*
Hastings†	22	25	28	32	A	V
Newhaven†	39	45	43	41	B	V
Huntshaw Cross	55	62	59	65	C/D	H
Keeleylang Hill	40	46	43	50	B	H
Knock More	33	26	23	29	A	H
Limavady	55	62	59	65	C/D	H
Londonderry†	51	44	41	47	B	V
Llanddona	57	63	60	53	C/D	H
Bethesda†	57	63	60	53	C/D	V
Betws-y-Coed†	21	27	24	31	A	V
Conway†	40	46	43	50	B	V
Mendip	58	64	61	54	C/D	H
Bath†	22	28	25	32	A	V
Bristol (Ilchester Cres.)†	40	46	43	50	B	V
Midhurst (West Sussex)	61	55	58	68	C/D	H
Moel-y-Parc	52	45	49	42	E	H
Oxford	57	63	60	53	C/D	H
Pontop Pike	58	64	61	54	C/D	H
Fenham†	21	27	24	31	A	V
Newton†	33	26	23	29	A	V
Weardale†	51	44	41	47	B	V
Presely	46	40	43	50	B	H
Redruth	51	44	41	47	B	H
Ridge Hill	22	28	25	32	A	H
Rosemarkie	39	45	49	42	B	H
Rosneath (Dunbartons).	58	64	61	54	C/D	H
Rowridge	31	24	27	21	A	H
Brighton†	57	63	60	53	C/D	V
Salisbury†	57	63	60	53	C/D	V
Ventnor†	39	45	49	42	B	V
Rumster Forest	31	27	24	21	A	H
Sandy Heath	31	27	24	21	A	H
Selkirk	55	62	59	65	C/D	H
Stockland Hill	33	26	23	29	A	H
Sudbury	51	44	41	47	B	H
Sutton Coldfield	46	40	43	50	B	H
Brierley Hill†	57	63	60	53	C/D	V
Bromsgrove†	21	27	24	31	A	V
Fenton (Stoke-on-Trent)†	21	27	24	31	A	V

Table 2.2. Band IV/V (UHF) stations and channels (*continued*)

Station	*Channels*					
	BBC1	*BBC2*	*ITV*	*Ch4*	*Aerial group*	*Polari- sation*
Kidderminster†	58	64	61	54	C/D	V
Lark Stoke†	33	26	23	29	A	V
Malvern†	56	62	66	68	C/D	V
Tacolneston	62	55	59	65	C/D	H
Aldeburgh†	33	26	23	30	A	V
West Runton†	33	26	23	29	A	V
Wrekin	26	33	23	29	A	H
Waltham	58	64	61	54	C/D	H
Wenvoe	44	51	41	47	B	H
Aberdare†	21	27	24	31	A	V
Bargoed†	21	27	24	31	A	V
Kilvey Hill†	33	26	23	29	A	V
Llanhilleth†	39	45	49	42	B	V
Maesteg†	22	28	25	32	A	V
Merthyr Tydfil†	22	28	25	32	A	V
Mynydd Machen†	33	26	23	29	A	V
Pontypool†	21	27	24	31	A	V
Pontypridd†	22	28	25	32	A	V
Rhondda†	33	26	23	29	A	V
Rhymney†	57	63	60	53	C/D	V
Winter Hill	55	62	59	65	C/D	H
Darwen†	39	45	49	42	B	V
Glossop†	22	28	25	32	A	V
Haslingden†	33	26	23	29	A	V
Kendal†	58	64	61	54	C/D	V
Lancaster†	31	27	24	21	A	V
Pendle Forest†	22	28	25	32	A	V
Saddleworth†	52	45	49	42	E	V
Skipton†	39	45	49	42	B	V
Todmorden†	39	45	49	42	B	V
Windermere†	51	44	41	47	B	V

Polarisation: H horizontal; V vertical
† Relay station.

broadcast purposes, and many accessories such as VCR machines and television games have their outputs on one of these three channels to avoid interference with broadcast signals. The sound and vision carrier frequencies for the UHF bands are given in *Table 2.1*.

As mentioned earlier, the four transmissions (BBC1, BBC2, ITV and Channel 4) are co-sited – i.e., they come from a common transmitter. The channels in a group are carefully chosen to avoid mutal interference, taking into account the design of the receiving aerial and the TV receiver itself. The transmitter network in the UK has been very carefully planned with regard to the power, positioning, and wavelength of each station, to minimise interference effects. In spite of this, at certain sites and particularly in certain atmospheric conditions, television reception can and does suffer from the effects of co-channel interference.

The number of transmitting-station sites in the UK runs to over five hundred, and it is not practical to detail them all here. *Table 2.2* indicates all main stations and the more significant relays. A full list of TV transmitting stations, and coverage maps if required, can be obtained from the BBC Engineering Information Service, Broadcasting House, London W1A 1AA, or the IBA Engineering Information Service, Crawley Court, Winchester, Southampton SO21 2QA.

3

Reception

The VHF and UHF radio waves which are used to carry the television sound and vision signals have a distinct characteristic of *propagation* and, unlike the medium-frequency radio waves used for ordinary AM sound broadcasting, behave similarly to light waves. This means that the waves exhibit the effects of reflection and diffraction (bending). Moreover, UHF signals can be considerably attenuated by obstructions in the line of travel of the waves. Indeed, towards the top end of the UHF bands (i.e., band V), the signals can be almost completely blocked by a large obstruction in their path.

Radio waves

The radio waves emitted by a transmitting aerial generally follow two different paths: the waves propagated along one path are called the *sky wave* and the other the *ground wave*. At medium freqencies, the sky wave is often bounced off the ionosphere to provide long-distance reception, while the ground wave satisfies, as far as possible, the demands of local reception. However, as the signal frequency rises, the ground wave more rapidly becomes attenuated due to the power losses of the earth. Thus on the short-wave bands, local reception soon fades out along with the ground wave, but long-range reception is sustained by the sky wave.

At very high and ultra-high frequencies, the ground wave disappears very near the transmitting aerial, but instead of there being a wave directed to the sky, the aerial system is arranged to deliver a *space wave* which tends to hug the surface of the earth. This is possible because the aerial system can be mounted on the top of a lofty mast, many wavelengths above earth. So it is the space wave which is used for television and other VHF and UHF services.

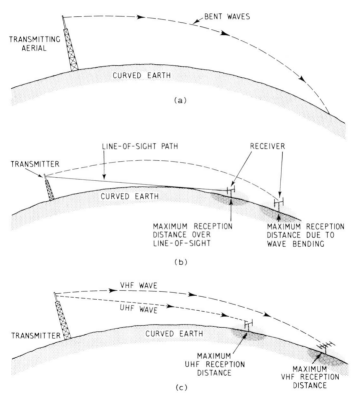

Figure 3.1. (a) Television transmission following the earth's curvature. (b) Bending gives greater reception distance than line-of-sight. (c) UHF undergoes less bending than VHF

55

High-frequency waves are held within the confines of the earth's atmosphere by the ionised layers surrounding it. These reflect them back to earth, and sometimes the earth causes further reflections back to the ionosphere, giving *skip-distance reception*. But VHF and UHF waves directed skywards often penetrate the ionosphere and disappear into space. Low- and medium-frequency signals thus employ the ground wave for medium-distance reception (i.e., regional reception), medium- and high-frequency (i.e., shortwave) signals the sky wave for long-distance reception, and VHF and UHF signals the space wave for relatively local reception only.

Television waves tend to hug the earth because they are diffracted round its curvature, as shown in *Figure 3.1(a)*. The waves thus fail to travel quite in straight lines, as do light waves. This diffraction makes possible television reception somewhat farther than the line-of-sight distance between the transmitting aerial and the receiving aerial, as indicated in *Figure 3.1(b)*.

Fringe and service areas

The amount of bending influences the maximum distance over which television signal can be usefully employed. Towards the limit of the bending, the signals become substantially weakened by the effects of normal attenuation (i.e., for the same reasons that light becomes dimmer as the distance of the source is increased from an observer); this is called the *fringe area*. The *service area* is essentially that within the line-of-sight distance of the transmitting and receiving aerials; it can be up to 30 or 40 km, depending on the heights of the aerials concerned. Aerial height can be quite important for good television reception.

The start of the fringe area is also influenced by the nature of the country over which the space wave is travelling. Hilly country affects the strength of the wave, and very high hills between the transmitter and the receiving aerial can cut down the signal strength a lot. They can also distort the signals and cause reflections; this problem is dealt with below.

The least attenuation and distortion occur over flat country, but large towns and built-up areas can modify the theoretically expected strength, even though the surrounding countryside may be flat. Towns also give rise to electrical interference fields that demand even stronger signals for their complete suppression.

Thus, in some locations the service area may extend up to 40 km or so, while in others, fringe-area effects may occur at little more than 10 km from a station or even less. Over good 'television country', acceptable reception may be possible up to 60 or 70 km from a high-power station.

As the frequency is increased, the degree of bending round the earth's surface is reduced, which means that the bending effect is less on the UHF channels than on VHF ones; this is the fundamental reason why the service area round a UHF station is less than that round a VHF station. The idea is illustrated in *Figure 3.1 (c)*.

Several other effects stem from the greater likeness of UHF signals to light waves. For instance, while a VHF signal may bend downwards slightly when passing over the top of a hill between the transmitter and receiver to give at least some signal at the receiving site, a UHF signal might not bend at all and thus be blocked wholly from the receiving site by the hill. Moreover, the much shorter wavelengths of the UHF signals compared with the VHF ones result in far greater trouble from signal reflection, for a radio wave can be reflected by an object no smaller than half its wavelength, and at UHF there are very many more such objects than at VHF. Also, UHF waves can be blocked completely by large man-made and natural objects. For total coverage of a country, therefore, many more UHF stations are required than VHF stations.

Shared channels
The local nature of television stations means that transmitters using a common or shared channel can be set up in different parts of the country without undue danger of interference.

Such shared working must be very carefully controlled, and only stations far from each other are given the same channel. As a further precaution, a high-powered main station is usually given one plane of signal polarisation, while a lower-powered secondary station working on the same channel is given the other plane of polarisation. Thus, the main station will have horizontal polarisation, needing horizontally mounted aerials, while the relay station usually has vertical polarisation, needing vertically mounted aerials.

In spite of these precautions, shared-channel (co-channel) interference does arise from time to time, particularly during early spring and settled summer weather conditions. The reason for this is that the waves undergo diffraction in the atmosphere directly above the earth, called the 'troposphere'. (This should not be confused with the ionosphere, which is a heavily ionised layer or layers *above* the earth's local atmosphere).

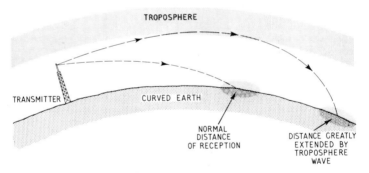

Figure 3.2. Tropospheric bending can extend the propagation of TV transmissions far beyond the normal reception range

Tropospheric diffraction affects mostly those space waves above the normal direct-route waves, and these waves are bent so that they 'illuminate' an area somewhat in advance of the fringe area, as shown in *Figure 3.2.* The bending effect is caused by the progressive fall in temperature of the earth's atmosphere with height above the earth, and the waves travelling through this environment are diffracted similarly to light waves passing from, say, air to glass or vice versa.

Tropospheric effects thus tend to propagate TV signals over distances far greater than are required for local television reception. So channels have to be shared – because of the limited bandspace available – not only within a country but also between countries, so that a transmission meant, say, for Scotland could appear in southern England on the local channel there or even on a local channel somewhere in Europe. Bad interference on the local transmission of the affected area thus results, and unfortunately there is not much that can be done to delete it. A typical co-channel interference effect is shown in *Figure 3.3*.

Figure 3.3. The effect of co-channel interference

It was stated earlier that high- and medium-frequency waves are bounced off the ionosphere for long-distance propagation. Sometimes, however, in certain layers of the ionosphere there is concentrated ion density (resulting from sunspot activity and certain electrical storms and effects from outer space), and these layers can reflect VHF waves as does the troposphere. Since the ionosphere is above the troposphere, television signals so propagated may appear many thousands of miles away from the transmitter, and reports of BBC1 signals on Channel 1 (VHF) as far distant as South Africa have been recorded. VHF signals are most affected.

Signal strengths

When a transmitter is working, the space round it becomes
'illuminated' by invisible radio waves, a *signal field* being
created. This can be measured with a field-strength meter
applied under very carefully controlled conditions, and a
field-strength contour map can be plotted by measuring at a
large number of points within the area that the transmitter is

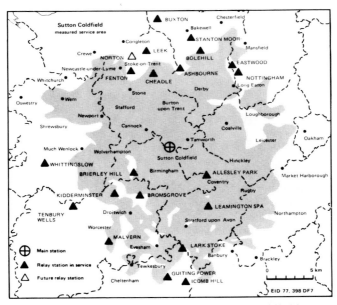

Figure 3.4. Signal-coverage map of the Sutton Coldfield UHF
transmitter. The service area is shown hatched

designed to serve. Such contour maps are published by the
television authorities for almost every station. The maps
reflect the nature of the topography in terms of lower signal
fields at the far side of hills, probably higher fields in valleys
and along rivers, and very low ones where large obstructions
occur in the signal line.

Figure 3.4 shows a contour map of the BBC's Sutton Coldfield station. The central cross represents the transmitter, and the hatched area is the measured service area. A number of triangles are present around the fringe area. These represent relay stations, many of which fill in 'shadow areas' in which reception from the main transmitter is unsatisfactory.

The amount of real signal abstracted from a signal field depends on the strength of the contour and on the gain of the aerial system. When an aerial is placed in a signal field, signal currents and voltages are induced into it and transferred to the receiver through the download (feeder or transmission line). If a signal-strength meter is connected at the end of the feeder instead of a receiver, the signal can be measured in actual voltage; an average signal is 1 mV.

The very strong signal field in the main service area usually means that an adequate signal will be extracted from the wave by the simplest of aerials, giving good reception quite easily. This implies, for UHF, a roof- or loft-mounted six- or eight-element array. In areas very close to the transmitter, a set-top aerial will often give sufficient signal strength, but performance is usually marred by multiple images ('ghosts') caused by signal reflections from the building and by objects (and people) in the room.

As the signal strength drops with increasing distance from the transmitter, larger and more elaborate aerial arrays are required. Where the signal strength is insufficient to produce, say, $300 \mu V$ at the receiver's aerial socket, noise will become apparent on the picture (see *Figure 7.16*). This is the familiar 'snowstorm' effect, and it corresponds to the hiss present on radio sound when receiving a distant station. In such cases, the use of a *masthead amplifier* or 'booster' will often improve reception. The masthead amplifier is a small solid-state wideband RF amplifier, designed to be mounted as close as possible to the dipole itself. So it is fitted with a weatherproof housing and is powered by a DC voltage coming up the feeder cable from a power unit at the receiver. It is necessary to mount the amplifier at the masthead because the coaxial feeder cable introduces losses of its own,

61

and a useful increase in signal-to-noise (S/N) ratio results from applying amplification before the signal is sent down the cable.

Noise

A signal strength, quoted in terms of voltage, means nothing by itself. Even if we have only 50 μV of signal arriving at the dipole, we can fairly easily apply sufficient amplification to raise this to 1 mV, which is the level normally aimed at for a UHF television signal. Nothing will be gained, however, because that 1 mV will consist largely of noise, and the picture may well be indiscernible among the snow. Similarly, a noise figure quoted by itself means little. In basic terms, 1 mV of noise would be quite acceptable if it were accompanied by 100 mV of signal. What is important is the signal-to-noise ratio. This is normally expressed in decibels: e.g., a S/N ratio of 20 dB implies a signal voltage 10 times greater than the noise voltage. Noise is generated by random movement of electrons in a conductor, and so it is being created at every stage in the system. Noise in a television picture becomes discernible as 'snow' or 'grain' when the S/N ratio falls below about 40 dB, representing a ratio of about 100:1. Because the noise generated in the receiving system is at a fixed level under full-gain conditions (more of this in Chapter 4), we have to maximise the RF signal level to achieve or exceed this figure. We cannot eliminate noise by amplifying it!

Receiving aerials

If a conductor is set up in a field of radiation from a transmitter, an induced RF current will flow in it. To extract the maximum energy from the transmitted signal, the aerial needs to be tuned to the frequency it is receiving; for television receiving aerials, a half-wave *dipole* is used. The distribution of voltage and current in a half-wave dipole is shown in *Figure 3.5*, in which it can be seen that the current is at a maximum, and voltage at a minimum, at the centre of the

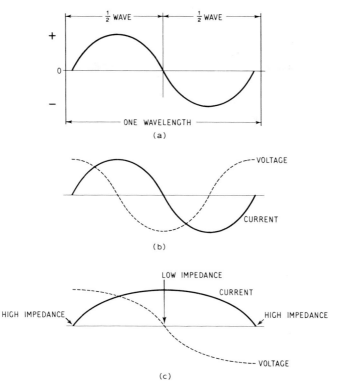

Figure 3.5. Voltage and current distribution. (a) A full signal cycle. (b) Current and voltage in a full-wave aerial, and (c) in a half-wave aerial

dipole. High current and low voltage correspond to low-impedance conditions, and so the lowest impedance point is at the dipole centre. It is a characteristic of this type of aerial that the impedance at the centre point is about $75\,\Omega$. We need to convey the signal to the television receiver with as little loss as possible, and to satisfy this requirement it is necessary that the impedances of the signal source (dipole), transmission system (feeder), and receiver (TV tuner) are matched, so that no energy is wasted. Coaxial cable is manufactured with a characteristic impedance of 75 Ω, so if we break the dipole at its centre point and connect the inner ends to the cable, we shall have an efficient transfer of power.

A single half-wave dipole would not make a very satisfactory TV receiving aerial. When mounted vertically, it has an omnidirectional characteristic, which means that it will pick up equally signals arriving from all directions. When mounted horizontally, its response takes on a figure-of-eight shape, so it offers some discrimination to signals arriving from different directions: it has 'directivity'. *Figure 3.6* shows these patterns. Another disadvantage of the single dipole is

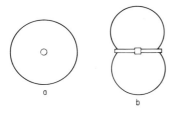

Figure 3.6. Polar diagrams for a single dipole. (a) Mounted vertically. (b) Mounted horizontally

that it is relatively insensitive,and a strong signal is necessary to induce a usable output from it. In fact, the single dipole is said to have 'unity gain' – i.e. O dB. It may be argued that an aerial cannot have gain, as it is not an amplifying device: true enough, but we shall see that a dipole can be made more sensitive by the addition of further elements.

The solution to the gain and directivity problems inherent in the single dipole is the addition of further elements to the aerial system. These are not connected to the feeder but spaced from the dipole along the boom which forms the 'backbone' of the aerial. Called 'parasitic' elements, they take the form of a reflector mounted behind the dipole (looking from the transmitter, that is) and a series of directors in front of it. This forms a 'Yagi array', named after its inventor, and shown in *Figure 3.7*. The reflector, as its name implies, reflects energy back to the dipole, and its spacing from the dipole is such that the reflected signal is in phase with that arriving at the dipole itself. The directors reinforce and concentrate the signal arriving at the dipole and have the effect of making the aerial more directional. This means that the aerial array becomes much more sensitive in the forward

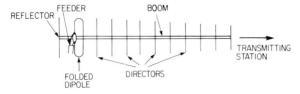

Figure 3.7. A Yagi array for use at UHF

direction at the expense of sensitivity at the sides and back (see *Figure 3.8*). This increased sensitivity accounts for the 'gain' of an aerial array, and the improvement offered by an 18-element Yagi array over a simple dipole is typically 14 dB.

This directivity of the aerial system is very important. It means that the aerial can be 'beamed' on to the required transmitter and its full gain realised in that direction only. Unwanted signals arriving off-beam are rejected, so there is a

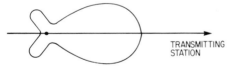

Figure 3.8. The polar diagram for a typical 18-element Yagi aerial

good degree of immunity from vehicle-ignition interference, co-channel transmissions, and reflected signals. Some examples of Yagi aerial systems are illustrated in *Figures 3.9* and *3.10*.

A quite different type of aerial is the *log-periodic* design. Although superficially similar to the Yagi aerial, it is totally different in principle. Instead of a boom, this aerial has a transmission line in the form of a pair of bus-bars runing the length of the array. A series of dipoles is connected to this

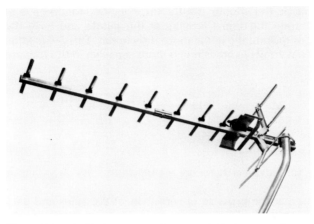

Figure 3.9. A 10-element roof-mounted Yagi array, shown mounted horizontally. It typifies an installation within the inner service aera of a main transmitter (Antiference Ltd)

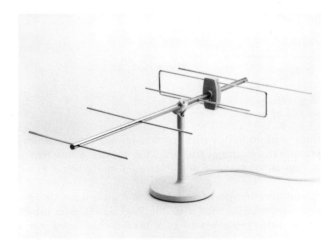

Figure 3.10. A set-top aerial, Yagi type (Antiference Ltd)

transmission line, their connections being transposed between adjacent dipoles. Starting with the longest dipole at one end, they become progressively shorter along the length of the aerial, thus giving the array its characteristic 'triangular'

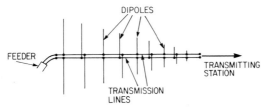

Figure 3.11. Principle of the log-periodic aerial. All elements are dipoles

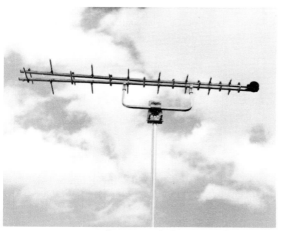

Figure 3.12. Roof-mounted log-periodic aerial (Jaybeam Ltd)

shape. The construction of this type of array is shown in *Figure 3.11*. Because each dipole comes into tune at its resonant frequency, the array can be made with very wideband response, but in gain and directivity it is far inferior to the Yagi type. It finds applications for TV reception where the channels of a particular group are widely spaced, and where reflections and co-channel signals are not troublesome. It is marketed in roof-mounting and set-top versions (see *Figure 3.12*).

Multiple images

Since television signals tend to be reflected from hills and other large objects, a television aerial may receive two or more signals, one the direct signal from the station and others reflected, as shown in *Figure 3.13*. This means that the set will display the main picture from the direct signal and a secondary or 'ghost' picture, received a very small fraction of a second later because the path taken by the reflected signal is longer. The ghost picture is displaced to the right of the main picture by an amount corresponding to the extra

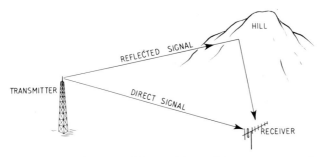

Figure 3.13. How a receiving aerial can pick up an unwanted reflected signal, received from the transmitter a little later than the direct signal

distance travelled by the reflected signal, and the ghost can be an ordinary positive picture, like the main one but less bright, or a negative one, with blacks white and whites black, as the result of phase-reversal effects occurring with the reflected signal. The extra path-length of the reflected signal can be worked out fairly simply from the horizontal scanning velocity of the spot, the velocity of travel of radio waves, and the distance of the ghost picture from the main one.

A number of reflective objects near by may give rise to multiple images, and the effect can then become serious, particularly in very hilly country. The use of a highly directional aerial can certainly help to solve the problem, as shown in *Figure 3.14*. At *(a)*, such an aerial is beamed for maximum pick-up of the direct signal, resulting in response A–B. The reflected signal is also giving a smaller but substantial response, A–C, and ghosting is apparent. Now, by reorientating the aerials in *(b)* so that the direct signal gives

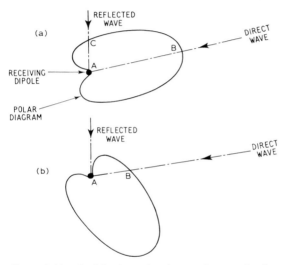

Figure 3.14. Aerial orientation for maximum rejection of interference or ghost images. (a) This gives response A-C to the reflected wave. (b) This gives zero response while retaining a usable response to the wanted signal

reduced response A–B, the reflected signal is completely eliminated since it arrives at a point on the aerial's *polar diagram* where the response is zero.

This technique can be usefully adopted in areas of interference, orientation of a directional aerial system not being made for maximum pick-up of the wanted signal but for *maximum discrimination* between the wanted signal and the interfering one.

Cable distribution systems

Direct reception of VHF and UHF signals via an aerial is not the only way of picking up TV pictures. The transmitting authorities use wideband cable links provided by British Telecom as well as microwave RF links for distributing their signals between studio and transmitter, and between studios and news gatherers. Distribution of TV programmes to domestic installations (*relay television*) via cable systems has the advantages of avoiding unsightly roof-mounted aerials, presenting 'clean' signals to all subscribers, and often offering a greater choice of programmes than could normally be available at the receiving site. Any future *cable television* system is likely to offer subscribers not only recorded programmes only available from their cable distributor in addition to local stations available off-air, but also out-of-region stations – thus a Scotsman in Cornwall could see Scottish television programmes.

The relay-cable operator will set up a master receiving installation at a carefully chosen site and ensure that his equipment is capable of producing impeccable signals on all transmissions to be used. A true relay system distributes the picture and sound signals in a form incompatible with ordinary TV receivers, and the receiving set is known as a 'terminal unit'. The sound signal is transmitted at AF, and to avoid losses in the resistance of the cables, the audio is distributed at high impedance, rather like the 100 V line system used in public-address equipment. Thus no power is required at the receiving end for sound reception, merely a volume control and matching transformer to suit the low-impedance loudspeaker.

The vision signal presents a greater problem. As we have seen, four or more programmes are required, each with a vision bandwidth of 5.5 MHz. The HF system of distribution is adopted, and this requires the use of a multicore cable with a separate twisted pair for each programme (see below). The video signal is amplitude-modulated on to an HF carrier, typically at 8.9 MHz. Vestigial sideband and negative modulation is used, so the spectrum of the relay signal is like that for broadcast television in *Figure 2.12 (b)*, with the important difference that either the upper or lower sideband may be used for relay television. With an HF carrier at 8.9 MHz, the lower sideband is transmitted. In some HF systems another, lower, carrier frequency of around 5.9 MHz is used, and in this case the upper sideband is transmitted.

Because of the stray capacitance and series resistance and inductance of the transmission cables in an HF relay system, the picture signal is vulnerable to various forms of distortion before it reaches the terminal unit. To combat this, *pre-emphasis* is applied to the signal before its launch into the cable system. This means that high frequencies are boosted, or emphasised, so that the frequency-dependent losses in

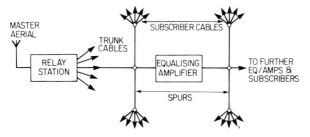

Figure 3.15. HF television distribution system

the cable are compensated for. At various points along the transmission line, amplification and equalisation are applied to maintain, as nearly as possible, an even signal and linear frequency response at each outlet. *Figure 3.15* shows the basis of the system for one vision channel.

Another problem inherent in multiwire HF distribution systems is *crosstalk*, sometimes known as 'crossview' in connection with television systems. This takes the form of an interfering pattern on the displayed picture and is due to pick-up of the vision signal from an unselected programme

71

on the network. This occurs because coupling takes place through the mutual capacitance between the individual wires in the multicore cable. There is also a transformer effect whereby the magnetic field surrounding each wire induces an EMF in a neighbouring conductor. These effects are overcome by the 'twisted pair' technique. The two involved with the transmission of any one programme are tightly twisted together, and this has the effect of concentrating their magnetic field into the space between them, minimising radiation from the pair. A further reduction in the nuisance value of crosstalk is achieved by locking the HF carriers together, so that any interference patterns are stationary, and sometimes by giving each HF programme carrier a phase offset from the others, the phase relationship being chosen to minimise the subjective effect of the interference pattern. At the receiving end, the HF signals are passed to the terminal unit.

The sound and vision signals are both distributed on the same pair, so simple filters are used to separate them. The vision signal then undergoes amplification and a degree of de-emphasis depending on the terminal unit's physical position in the network. The vision signal processing is the same as for a conventional receiver, which we will meet in the next chapter.

Fibre-optic cable transmission
We have seen that propagation of television signals along conventional cables is fraught with problems. One solution to these problems is the use of fibre-optic cables. A system already in use for sending colour television signals along a fibre-optic line uses for each programme a single fibre of 125 microns diameter, into which the signal is launched by an LED (light-emitting diode). The LED is operated in class A at a power of about 100 mW, and its light output is modulated with the same HF AM system as described above for use in cables. The glass-fibre line has a transmission characteristic which is almost perfect, and this is virtually unaffected by such physical constraints as coiling the line or passing it round corners.

There is a school of thought which says that the use of the 'ether' for broadcasting purposes is wasteful, since both the sender and receiver are static and quite capable of being wired to each other. This is a valid point of view, and it conjures up visions of radio and TV programmes being piped into every dwelling along with gas, telephone, and electricity services. It would have the advantage of releasing more broadcasting-spectrum space to those mobile services which cannot be wired up, and it has the further attraction of offering a 'talkback' facility from the subscriber to the broadcaster or other service. However, with over 1000 transmitting sites and many millions of TV aerials established and in use, this is necessarily a long-term policy, although the growing popularity of Pay-TV, in which subscribers buy their programmes by the hour, may bring this idea to fruition alongside conventional broadcasting techniques.

Communal aerial systems

The need often arises for a distribution system on a much smaller scale than that described above for a full-blown relay system. The simplest example is that of a dwelling where two or more receivers are required to work from a single aerial. In TV showrooms and workshops, many TV outlet points are required, and the same is true of hotels and blocks of flats; it is not practical or economic to provide a separate aerial system for each receiver in use. Sometimes in these circumstances, set-top aerials are resorted to, usually giving poor results. Again, a small group of dwellings may lie in the shadow, signalwise, of a hill, and need to be served by a hill-mounted receiving aerial with a feeder down to a convenient distrubition point near the houses (see *Figure 3.16*). Again, sometimes local authorities are reluctant to grant planning permission for individual roof-mounted aerials, particularly where the property is owned by the authority.

In any but the simplest case (two receivers being fed from one aerial in a good reception area), amplification is required to overcome the losses in the necessary cables, splitters, and outlets. Wherever signal splitting is necessary, it is essential to maintain correct impedance (invariably 75 Ω) at all points,

73

and for this purpose simple resistive or inductive splitters are used. A simple passive two-way splitter, which may also be used as a combiner, is shown in *Figure 3.17*.

Where a large number of outlets is required, a distribution amplifier will be necessary. This is a wideband transistorised amplifier, usually covering the whole of UHF bands IV and V

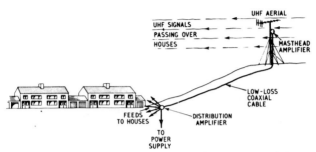

Figure 3.16. A communal UHF aerial system used in a 'shadow' area. The shared aerial is mounted on the hilltop, and the signals are distributed to the screened premises

and mains-powered. Its output is sufficiently high to drive, typically, 12 outlet points via either an external or a built-in splitter unit. Such an amplifier is illustrated in *Figure 3.18*, along with a six-way passive splitter. The gain of the unit is such that the output available at each outlet is comparable with the RF signal input level to the amplifier. For a 12-outlet system, this implies a gain of 12 times or about 22 dB.

Figure 3.17. A two-way passive splitter/combiner (RS Components Ltd)

Where more outlets are required than are available from a single distribution amplifier, a second or further amplifiers can be used, each fed from one of the outputs of the first. In these 'chain' applications (*Figure 3.19*), it is very important

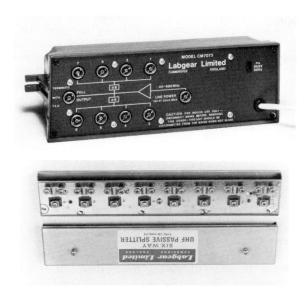

Figure 3.18. A typical distribution amplifier and six-way passive splitter (Labgear Ltd)

that the system is carefully engineered to avoid amplifier and receiver overloading and consequent cross-modulation effects.

Linearity and cross-modulation

In all the transmission systems we have been describing, it is essential that amplifying systems have good linearity. This means that a graph of input voltage plotted against output voltage should be a straight line, and that the output of an amplifier must be a faithful copy of the original signal. Where an amplifier has poor linearity (either because of bad design or, much more often, because too great an input signal is driving it beyond the linear region of its operation), the various components of its input signal 'jangle' together to produce *inter-modulation* and cross-modulation. The former produces 'beat' signals on the sum and difference frequencies of the components of the signal being handled (i.e., sound carrier, vision carrier, and chrominance sidebands of the wanted television channel). Cross-modulation, on the

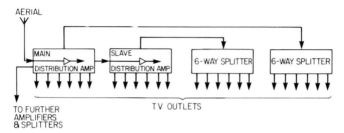

Figure 3.19. An amplified distribution system with multiple outlets. VHF radio and TV can be distributed simultaneously on the same network if required

other hand, has the effect of 'mixing' the wanted carrier with an adjacent unwanted carrier, giving rise to *adjacent-channel interference*. All these effects result in spurious patterns on the display, and once introduced into a system, whether by overload of a masthead amplifier or too high a signal level into a tuner, cannot be eliminated no matter how linear the following stages may be. One common effect of overload and resulting non-linearity is the 'vision buzz' on TV sound, which manifests itself as a rough buzz over the sound, which varies in sympathy with the picture content.

Direct satellite broadcasting

Satellite links have for many years been used for transmission of television signals between continents. For domestic viewers, a more tangible form of satellite reception is planned to

76

start in 1986, in which a geostationary (synchronous) satellite is used to beam SHF (super-high-frequency – above 3000 MHz) signals direct to the viewer. The frequencies allocated are in the range 11.7 GHz to 12.5 GHz. (One gigahertz, GHz, is equal to 1000 MHz.) A parabolic dish aerial will be required; provided it can 'see' the correct part of the sky, roof mounting may not be necessary. A frequency converter will be necessary to convert the incoming signals into a form suitable for the conventional television receiver. The only problem will be the cost of the dish aerial and conversion equipment, which must cost substantially more than a conventional UHF aerial system.

4

The receiver – signal processing

In previous chapters, we have seen how the television image is analysed, scanned, and turned into an electrical signal, and how syncs and sound are added before the signal is launched into the transmission medium, be it space or a cable system. The job of the television receiver is to unscramble this composite signal and present it as a picture on the screen and as sound from the loudspeaker. In this and the next chapter, we shall see how this is done and follow the signal from the aerial socket to the picture tube and loudspeaker.

Receiver outline

A block diagram of a television receiver is shown in *Figure 4.1*. The UHF signal from the aerial first enters the tuner, whose function it is to select the wanted channel from the many others present at the aerial, amplify it somewhat, and convert its freqeuncy to a lower one more suitable for handling within the set. The signal then passes into the IF (*intermediate frequency*) amplifier, which is a tuned amplifier. This rejects unwanted signals present in the tuner output and raises the level of the wanted signal to a point where it is suitable for *detection* or demodulation. At this point, the sound and vision signals split, the sound signal going to its own amplifier and detector to recreate the original audio (AF) signal. This is amplified and passed to the loudspeaker.

The vision signal, in composite (sync pulses present) form, now undergoes further amplification in the video amplifier and is finally passed to the picture tube to modulate the beam, so that the picture can be built up on the tube screen. At a point before the picture tube, the video signal is 'tapped off' and applied to the synchronising separator ('sync separator' for short) which strips the vision signal away from the sync pulses. These pulses are then processed by an integrator (for the field sync) and a differentiator (line sync) to convert them into a form suitable for synchronising, or holding in step, the timebase oscillators.

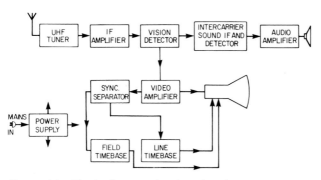

Figure 4.1. Block diagram showing the elements of a TV receiver

Each complete timebase (field and line) consists of an oscillator to generate the scanning waveforms and a power output stage to drive the scan coils. The line timebase has a more sophisticated synchronising system than the field timebase, and its output stage is pressed into service to generate the high-voltage supplies for the picture tube. The receiver also needs a power-supply circuit to convert the incoming mains to smooth DC at a voltage suitable for powering the amplifiers and timebases.

UHF tuners

Television tuners fall into two categories – mechanical, and electronic or 'varicap' (variable capacitance). The names describe the method of tuning, or selecting the required channel. Mechanical tuners are used in some portable sets and may be encountered in some old receivers. In these, a shaft runs through the tuner: fitted to it is a series of metal vanes, which mesh with stationary vanes fixed to the tuner body. Tuning is accomplished in the traditional way by rotating the shaft to vary the capacitance in the LC tuned circuits of the tuner. Much more common, and more flexible, is the varicap tuner, an example of which is illustrated in *Figure 4.2*. In place of the shaft and vanes, the required

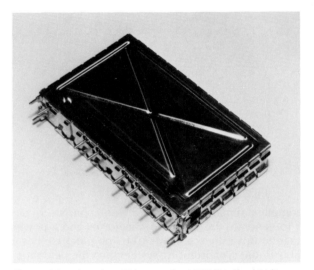

Figure 4.2. A varicap TV tuner for UHF (Mullard Ltd)

capacitance swing is achieved by the use of special varicap diodes. In any diode, when the junction is reverse-biased, a depletion layer exists between the anode and cathode regions of the device. This depletion layer varies in thickness with the applied bias and represents the dielectric of a capacitor, whose plates are formed by the diode's anode and cathode. Thus the effective capacitance of the device depends on the degree of reverse bias applied. Varicap diodes are designed to exploit this effect, and for TV tuner applications, a voltage variation of 0 V–30 V is sufficient to swing their capacitance far enough – when used in conjunction with suitable inductors – to cover the entire UHF television band. VHF tuners are also available in varicap form. The use of this type of tuner allows great flexibility in receiver design, because the tuning is carried out by varying a DC voltage, the tuner need not be mechanically connected to the programme selector system, and remote control of channel tuning is easily arranged.

A circuit diagram of a varicap tuner is given in *Figure 4.3*. The UHF signal is applied to the emitter of the first RF amplifier *VT1*. This stage is operated in 'common base' configuration, meaning that the base is grounded signal-wise, the signal being applied to the emitter and extracted at the collector. This arrangement gives low-noise amplification, and provides good isolation between the input and output electrodes of the transistor. At UHF, this is very important. The transistor is designed in such a way that its gain can be varied by the applied base current for AGC (automatic gain control) purposes: we shall see why later. The signal now passes into the second screened box in the tuner. This screening is essential to prevent capacitive coupling and thus prevents feedback and instability in the tuner. In the second box, the required channel is tuned by the resonant circuit formed by *L5/6/7* and the varicap diode *D1*. The selected signal is passed into the emitter of *VT2*, which works in the same fashion as *VT1*, and further amplifies the signal. *VT2*'s output is developed across a further series of tuned circuits, *L9* to *L15*.

Television tuners are superhet devices: i.e., they change the frequency of the incoming signal by 'beating' its carrier with a locally-generated signal to produce a *difference* frequency which is the wanted IF. In our tuner, *VT3* is the local oscillator, and this runs at a frequency above the incoming signal. Let us assume that we are tuned to channel 40, whose vision signal is radiated at 623.25 MHz. The applied tuning voltage would be such that the varicap diodes *D1/2/3* were holding their tuned circuits resonant at this frequency, thus

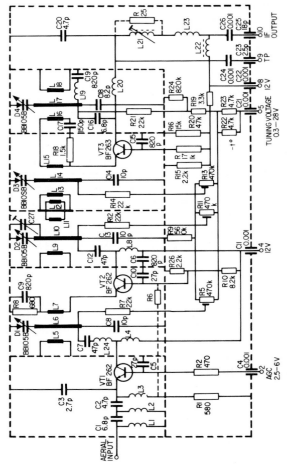

Figure 4.3. Internal circuit of a varicap tuner (Mullard Ltd)

82

offering high gain by VT1 and VT2 to channel 40. The same tuning voltage applied to varicap D4 is arranged to hold its tuned circuit L16/17/18 resonant at 662.75 MHz, and because of the deliberate feedback from VT3 collector circuit via L15 to its emitter, the transistor oscillates at this frequency. Thus the emitter of VT3 is furnished with two signals, the amplified incoming 623.25 MHz vision carrier from L16, and the local oscillator signal at 662.75 MHz. These signals beat together, so that sum and difference frequencies are produced. The sum frequency is so high that it is lost, but the difference frequency (662.75–623.25 = 39.5 MHz) appears at the collector of VT3. The oscillator signal is filtered out by L20, and the vision carrier, now at 39.5 MHz, is selected by L21 and passed out of the tuner via L23.

It will be remembered that the transmitter radiates on the vestigial-sideband system, and that the upper sideband is the one which is broadcast. Because the local oscillator frequency is 'high' with respect to the transmitter, the upper sideband is nearer the local oscillator frequency and thus produces a smaller 'difference' frequency. This means that, after

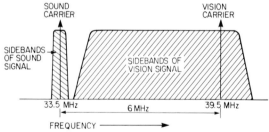

Figure 4.4. Spectrum of the IF signal as it emerges from the tuner

conversion in the tuner to IF, the sideband will appear below the vision carrier frequency. In fact the carrier spectrum will be a mirror image of the transmitted signal, as shown in *Figure 4.4.* As can be seen, the sound signal, transmitted for channel 40 on 629.25 MHz, now appears at 33.5 MHz, this being the beat or difference frequency between the sound carrier and the local oscillator (662.75–629.25 = 33.5 MHz).

It is essential that all the tuned circuits in the tuner keep in step with one another, so that when receiving any one channel, the tuned circuits controlled by *D1, D2* and *D3* are all exactly resonant on that channel, and that this occurs all the way up from channel 21 to channel 68 (in UHF tuners, that is). Likewise, the oscillator frequency, controlled by *D4*, must always be precisely 39.5 MHz above the incoming frequency, and this spacing must be maintained throughout the band. If these conditions are not realised, the result will be impaired signal-to-noise ratio, causing grainy pictures, as explained in Chapter 3. This tracking procedure is no mean feat and is achieved by using matched sets of varicap diodes in manufacture and by careful factory trimming of the tracking controls *R5/11/13*.

Much of the noise we see when viewing a receiver working on a weak signal is generated within the tuner itself. To minimise noise, the input circuit *L1/2, C1/2* of the tuner is broadband, and matched to the feeder impedance of 75 Ω. Under normal conditions, *VT1* is operated at full gain for optimum noise performance. TV receivers have to work with very wide variations of signal strength, however, and a typical requirement is that the set should be capable of operating with an RF input signal ranging from about 100 μV to several millivolts. At the lower extreme, picture noise will obviously be present, but at high levels of input (say above 3 mV), there is the risk that overloading will occur in the tuner's RF amplifiers, with consequent cross- and inter-modulation. At such a signal level, tuner noise is not significant, so we can afford to turn down the gain of *VT1* (and in some tuners, the second amplifier as well) to prevent non-linearity. Thus tuner gain is reduced for high signal strengths by increasing the voltage at pin 2. This has the effect of increasing the current in *VT1*, whose gain falls in these circumstances. Some tuners achieve this AGC (automatic gain control) action by the use of special diode attenuators at the tuner input. Where the RF signal level is very high (say 10 mV upwards), an external attenuator may be necessary at the aerial socket to reduce the signal to a level with which the tuner can cope.

To sum up, the television tuner is a remarkable device! It

will deliver sound and vision signals at fixed frequencies for any channels throughout the UHF bands (and VHF bands, if the set is a multi-band type), and cope with a wide variation of signal strength. It amplifies incoming signals, with the minimum of noise, to the point where the noise figure of succeeding stages is not a problem. It is simply tuned by application of a selected voltage to its varicap diode system, and it lends itself to AGC and AFC (automatic frequency control). The channel tuning is proportional to applied voltage, so this voltage source must be stable and jitter-free, or tuning drift will occur.

Tuning-voltage sources

In simple receivers, particularly portable sets, the tuning-voltage source may be as simple as a single potentiometer with a tuning knob calibrated in channel numbers. More often, a push-button system is used, in which a series of push-buttons is provided, labelled 'BBC1', 'BBC2', 'ITV1', and so on. Each selects the output of a pre-set potentiometer, arranged so that each potentiometer ('pot') taps off just the right tuning voltage to correspond to the wanted channel. Whatever the channel-selection system, the pot network is provided with a potential of about 33 V, derived from a special IC (*integrated circuit – 'chip'*) designed for the purpose. It is a voltage stabiliser with a zener-like characteristic but an extremely low *temperature coefficient*, so that its voltage is very stable with time and temperature.

Some sets are fitted with 'touch-tuning', in which the touch of a fingertip on a pair of contacts selects the required programme. This system operates by amplifying the very small current through the touch-pads via the finger to the point where they can close an electronic switch, in the form of a transistor or IC, to select the pre-tuned potentiometer. Remote-control systems of channel selection do likewise; we shall meet them in more detail later.

A quite different tuning system is used in some receivers. It is known as *sweep-tuning*, or 'self-seek', and sweeps the tuner through the whole band electronically. When the sweep is initiated by the viewer's 'seek' control, a gradually-increasing voltage is generated within the set and rises from zero towards 30 V. This is applied to the tuning pin of the tuner and takes the tuner up the TV band, starting at channel 21 for UHF. When a television signal is encountered, an IF output is produced, and the set recognises this as such. The sweep voltage is then 'frozen' so that the set remains tuned to that signal. If required, the channel for that signal can be committed to the memory, a device which stores the tuning voltage corresponding to that channel. Alternatively, the sweep can be continued by the viewer until the required station is found. Once all required channels have been 'stored' in this way, they can be recovered from the memory at will, appearing in the form of a voltage corresponding to the channel required, which when applied to the tuner will bring up the required programme (see *Figure 7.9*). Usually, the sweep process is simulated by a display on screen, taking the form of a horizontal or vertical line moving along a scale calibrated in channel numbers.

Although the electronics to accomplish this are complex, the system is made practical by the availability of purpose-designed ICs, which take up little space or energy.

IF amplifier

IF stands for 'intermediate frequency', so called because it is below the transmitter frequency but above the frequency of the video and sound signals themselves. For various reasons concerned with interference, amplifier design, and detector requirements, the vision IF in the UK has been standardised at 39.5 MHz. This means, because the sound is transmitted 6 MHz above the vision carrier, that the sound IF will be 33.5 MHz. The spectrum of signals presented to the IF amplifier (often known, for historical reasons, as the 'IF strip') is shown in *Figure 4.4*. This however, is drawn for a single channel. Again taking channel 40 as our example, the signal being transmitted on channel 40 is what appears in that diagram. What happens if signals on channels 39 or 41 are present at the tuner input? The tuner is not sufficiently *selective* to reject them, so they will appear as spurious out-of-band signals on each side of the wanted channel (see *Figure 4.5*). These signals must not reach the vision detector, or patterns and interference will mar the picture. Clearly, the IF amplifier must offer not only gain, but selectivity. It must reject out-of-band signals and selectively amplify those signals corresponding to sound and vision of the required channel so that they appear in the correct proportions at the detector.

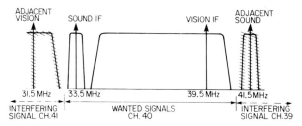

Figure 4.5. IF signal spectrum with adjacent channel signals present

If we were to plot a graph of gain against frequency for the IF amplifier, it would look similar to *Figure 4.6*. This is known as the 'IF response curve', and we can see that a deep notch appears at each side of the passband. These notches correspond to adjacent vision carrier (31.5 MHz) and adjacent sound carrier (41.5 MHz). They effectively 'kill' any interfering signals from the channels above and below the wanted one. The shape of the response curve itself is puzzling, though. Why not an equal response to all signals within the passband? In our description of the transmitter, we showed that it is not possible to wholly eliminate the unwanted sideband from the broadcast signal, hence the term 'vestigial sideband signal'. This sideband is still present in the IF signal, and if we amplified the whole IF spectrum equally, for lower video frequencies both sidebands would appear at the vision detector at equal strength. This would result in twice the signal energy being present for low frequencies (up to 1.25 MHz), so the detector output would be higher for these

LF video signals, giving undue emphasis to them in the reproduced picture. To balance this out, the IF amplifier is given a falling response about the vision IF of 39.5 MHz. The point selected, 6 dB below full gain, gives a 'flat' response at the detector.

There is also a *notch* in the response curve at the sound carrier frequency, 33.5 MHz. As we shall see, the FM sound carrier undergoes a double superhet process in the receiver, and to achieve this it is necessary that the sound carrier

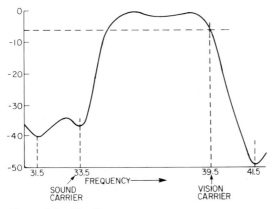

Figure 4.6. The IF response curve

Figure 4.7. SAW filter for IF response shaping (actual size)

passes through the IF amplifier at a lower level than the vision signal. This is achieved by depressing the level of the sound carrier at 33.5 MHz to a point about 35 dB below full gain.

The signal which emerges from the tuner, then, is first passed into a shaping circuit to give the characteristic response curve required. In older designs of receiver, this took

88

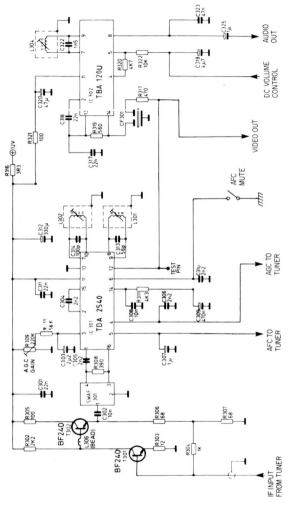

Figure 4.8. IF amplifier using SAW filter and ICs (ITT Consumer Products Ltd)

89

the form of a series of tuned circuits in discrete (separate, individual-component) form. These were made up of inductors and capacitors in the shape of filters which formed the required response curve but which had to be carefully tuned individually with special instruments. Current receivers use a SAW (surface acoustic wave) filter, a small piezoelectric device, for IF response shaping. The device is pictured in *Figure 4.7*. Its internal construction takes the form of a slice of crystal (typically, lithium niobate) with input and output *transducers* mounted at opposite ends. The input transducer converts the IF signal into an accoustic (mechanical) wave which travels over the surface of the crystal to the output transducer. Because of the construction of the input transducer, some frequencies are attenuated, and the response of the device is 'tailored' in manufacture by precision design of the transducers. The signal, having been propagated across the device, is converted back from mechanical to electrical form by the output transducer.

The selectivity circuit, then, is followed by amplification. If the IF amplifier is in discrete form, three transistor amplifier stages are used, but more commonly an IC is employed. Depending on the type of detector used, the IF amplifier needs a gain capability of up to 80 dB to cope with the variations in signal strength which may be encountered. A typical IF amplifier using a SAW filter is shown in *Figure 4.8*. The IF output from the tuner is first applied to a two-stage fixed-gain wideband amplifier *T301, T302*. This raises the IF signal level to compensate for the *insertion loss* of the SAW filter, typically 20 dB. The output is taken from the emitter of *T302*, which has a low impedance to provide optimum working conditions for the filter. The output from the SAW filter is developed across *R308* and passed into the IF amplifying chip *IC301* on pins *1* and *16*, via the DC blocking capacitor *C305*. The IC (in this case, type TDA 2540) provides several other functions than gain, and we will examine these in turn.

Automatic gain control
The gain or amplification of the IF amplifier determines the level of video drive presented to the picture tube, and hence

90

the contrast of the picture. It is important that this does not vary between channels or with varying signal strengths, so an AGC system is necessary. This is provided within the chip in the following way.

The amplitude of the signal coming from the tuner is measured by sampling the signal at a point where it is at a constant level regardless of picture content – i.e., during the sync or blanking period. This gives a true indication of signal level, and a control voltage is produced by this sampling method which is proportional to signal level; the control voltage is now applied to the first amplifying stage within the chip in such a way that its gain is increased when signal level falls, and decreased when signal level rises. This has the effect of 'ironing out' variations in signal level, so that at all usable RF signal levels, a constant-amplitude output is presented to the vision detector and finally to the picture tube.

We have already described how very high RF signal levels can overload the tuner and cause undesirable effects. At very high signal levels, the IF amplifier has been biassed back to minimum-gain conditions, and at a point set by the AGC crossover control *R309*, the voltage on pin *4* of the chip starts to rise. This voltage is passed to the AGC control pin on the tuner, and as the RF input signal rises, the gain of the tuner is progressively reduced.

Automatic frequency control
Although the frequency of the transmitter is closely controlled and cannot vary, and in spite of the care taken in tuner manufacture to minimise drift, the effects of temperature and ageing on the oscillator stage in the tuner are such that regular re-tuning of the receiver would be necessary. To overcome this, a further function of the IF chip is to provide an AFC output to compensate for tuner drift.

When the receiver is correctly tuned, the vision IF carrier is precisely 39.5 MHz. If tuning drift occurs, the vision carrier moves away from this frequency. Associated with the IF amplifier chip is a tuned circuit *L302/C314*, stable and resonant at 39.5 MHz. The incoming vision IF frequency is compared with this reference, and any deviation due to tuning drift gives rise to an *error voltage* which appears at pin 5 of the chip. When the set is correctly tuned, this voltage sits at a nominal reference point and is made to move upwards or downwards from that point, depending on the amount and direction of the drift. Let us assume that the oscillator frequency in the tuner rises, perhaps as a result of an increase in ambient temperature. This will increase the difference between local oscillator and transmitted carrier frequencies, so the vision IF will move up from 39.5 to, say, 40 MHz. When compared with the reference *L302/C314*, this will give rise to a downward swing in the AFC voltage at pin 5 of the chip. This reduction in voltage is applied (with the tuning voltage itself) to the tuning pin of the varicap tuner, so that it corrects oscillator frequency, thus pulling the receiver back into tune.

If the AFC were in operation when the set was being tuned, it would mask the correct tuning point and make it difficult to set up the tuning pots correctly. Most manufacturers arrange to cancel or *mute* the AFC circuit when tuning is being carried out; in *Figure 4.8* this is simply done within the chip when its pin 6 is grounded. Pin 6, then, is taken to a switch on the cover flap of the tuning pots so that the AFC is muted whenever the flap is opened for tuning adjustment.

Vision detector
Vision detection is the final process in the IF circuit; it consists of removing the vision signal from the vehicle which has carried it all the way from the transmitter – the vision carrier. Because the vision signal is amplitude-modulated on to the carrier at the transmitter, all that is required is a simple diode detector, similar to that used in AM radio receivers. This is known as an *envelope detector* and works by rectifying the carrier and charging a capacitor to the peak amplitude of the carrier wave. Thus the voltage across the capacitor is proportional to the modulation depth and so is an image of the modulating waveform, here the video signal. After detection, a degree of filtering is carried out to remove any

residual carrier signal, and the video signal is now in the same form as was present in the studio.

Because the vision detector is not completely linear, cross modulation will take place when more than one signal is present at its input. It will be recalled that the frequency-modulated sound signal at 33.5 MHz is passed through the IF amplifier at low level, and when this appears at the vision detector, it beats with the vision carrier at 39.5 MHz to give sum and difference frequencies. The sum frequency, 73 MHz, is immediately lost, but the difference beat, 39.5–33.5 = 6 MHz, appears alongside the video signal at the detector output. Because the sound signal is frequency-modulated, the 6 MHz beat frequency is similarly frequency-modulated, so that the broadcast sound signal can later be extracted from this 6 MHz carrier. It is now clear why we inserted the 'notch' in the IF response curve at 33.5 MHz. For this beat system to work, the sound signal must arrive at the vision detector at a level lower than any possible negative excursion of the vision signal, otherwise the sound signal will have vision components impressed on it, and this will give rise to a coarse buzz or crackle on sound.

Synchronous detection
There is an alternative to the simple diode envelope detector. It is known as a 'synchronous detector' and is embodied in ICs for television IF purposes such as the one in *Figure 4.8*. Basically, the arrangement may consist of a *bridge* of four diodes (see *Figure 4.9*). The incoming amplitude-modulated signal is applied to the bridge input and in the absence of any reference drive, none of the diodes conduct, so no output is produced. Imagine that the input consists of an AM telvision signal at 39.5 MHz. If we apply a reference signal to the upper and lower points on the bridge, all four diodes will conduct on each positive half-cycle at *A*, thus shorting together all terminals of the diode bridge. Now, if we arrange that the reference signal is at exactly the same frequency as the incoming carrier – i.e. 39.5 MHz – and that the input and reference signals are in phase (coincident in time), the bridge will conduct on each positive peak of the incoming carrier.

Thus the capacitor *C1* is connected across the carrier source for an instant on each positive half-cycle, and it will charge to the peak level of each. The load resistor *R1* will start to discharge the capacitor between pulses, so the average charge on *C1* will represent the instantaneous amplitude of the carrier wave. Because the carrier is amplitude-modulated, the modulation signal (vision signal) appears across *R1*. The reference signal is derived from the IF carrier

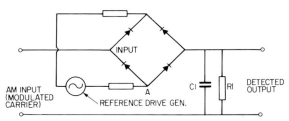

Figure 4.9.   Diode-bridge type of synchronous detector

itself and in our example (*Figure 4.8* again) is developed across the tuned circuit *L301/C313* and internally applied to the synchronous detector within the chip. The video output, then, appears at pin *12* of the chip, ready for further amplification before it is fed to the picture tube.

Video amplifier

A typical video detector provides an output between 2 V and 4 V amplitude, depending on type. The picture tube, however, requires a voltage swing of, say, 70 V to drive it from black (tube cutoff) to peak white. We need an amplifier capable of providing this voltage swing over the full video bandwidth of 5.5 MHz if we are to see all the detail in the picture.

A typical video amplifier is shown in *Figure 4.10*. First comes a stage to match the detector output impedance to the lower input impedance of the second stage. This is *TR1*, in a

circuit known as an 'emitter-follower'. This has the characteristics of high input impedance, zero voltage gain, and high
power gain. Its output is developed across its emitter resistor
R1, and here is inserted a gain control with which picture
contrast can be adjusted. The selected level of video signal is
tapped off at the slider of the contrast control *VR1* and
applied to the base of the video output transistor *TR2*. This is
a voltage amplifier with a gain of about 25, working from a
necessarily high supply voltage of, say, 150 V. Its output
appears across its collector load resistor *R4* and is applied to
the cathode of the picture tube. The fact that the synchronising pulses are still present does not have any effect on the

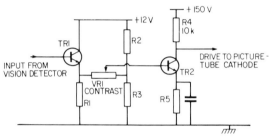

Figure 4.10. Typical two-stage video amplifier

picture display because, when the brightness is correctly
adjusted, the black level of the picture signal drives the tube
just to cutoff. Any further reduction of the video signal, such
as that imposed by the sync pulses, merely drives the tube
further into cutoff – i.e., 'blacker-than-black' – and no sign of
it is seen on the display.

At first sight, one may expect to see the picture tube driven
at its grid with the video signal. The important thing, though,
is that the beam intensity is modulated by the bias between
the cathode and grid, so that if the tube's grid is held at a
constant potential, the picture tube will work quite happily
with cathode modulation. It is often more convenient in
practical receivers to modulate the tube cathode, though the
the practice is not universal. It will be recalled that to

intensify the electron beam in the picture tube, the grid voltage must move in a positive direction, so to achieve the same effect with cathode drive, the video-signal must move negatively for white and in a positive direction for grey and black signals. This is convenient, because our video amplifier *TR2* inverts the applied signal (*common-emitter* configuration), thus producing correct cathode drive from a positive-for-white detector output.

The video signal at the collector of *TR2* is at fairly high impedance ($<10\,K\Omega$, the value of the collector load resistor) and any capacitance here will tend to shunt away the high-frequency components of the video signal, causing a loss of definition. A certain capacitance effect is present in the transistor itself, and stray capacitance appears in the collector wiring, tube base connector, and tube cathode itself. A convincing demonstration of this effect can be made by placing a finger on the cathode lead of the picture tube when the set is operating (make sure that no other conductive surface is touched at the same time). The introduction of the large capacitance of the human body has the effect of killing all the fine detail in the picture.

Tube ancillary circuits

As *Figure 4.11* shows, several other services are required by the picture tube itself. As we have just seen, the video signal drives the picture tube from black to white, corresponding to cutoff and full-beam current respectively. It is necessary to set up the bias on the picture tube so that tube cutoff occurs just at black level, and so the DC voltage on the control grid is variable by a potentiometer, *VR1*, which forms the brightness control. Also fed to either the grid or the cathode are blanking pulses, in *Figure 4.11* to the grid. What are they for? In earlier chapters, we have seen how the deflection circuits cause the spot to fly back to its starting point at the end of each line- and field-scanning stroke. This process cannot be achieved instantaneously, and during the time it is taking place, the spot is 'at large' on the screen, albeit moving very rapidly. Although the speed of the flyback tends to reduce

96

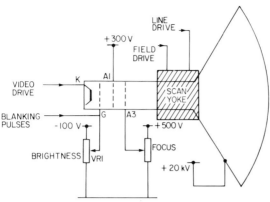

Figure 4.11. The services required by the picture tube:
four waveforms and four levels of DC voltage

the visibility of the spot, if steps are not taken to blackout the
spot during flyback, spurious lines and patterns appear, such
as those shown in *Figure 4.12*, which illustrates the effect of
unsuppressed field flyback. Blackout pulses, derived from
the field and line timebases (explained in the next chapter)
are therefore necessary. This completes the list of modulating signals to the picture-tube electrodes. The other electrodes in the tube are furnished with fixed potentials; we
shall see in the next chapter how these are derived.

Figure 4.12. Lack of blanking pulses –
the effect of unsuppressed field flybacks

Sound stages

Let us now trace the career of the sound signal. It was said earlier that the sound signal undergoes a double-superhet process in the receiver, and (taking again the example of channel 40), sound is radiated on a carrier of 629.25 MHz. On emergence from the tuner, we saw that the sound signal was translated to 33.5 MHz, still frequency-modulated, and passed through the IF strip in this form until it encountered the vision detector. Here the second heterodyne process took place, and we left the sound signal in the form of a low-level 6 MHz carrier at the vision detector output. This is known as the *intercarrier* process because the sound carrier is now at a second intermediate frequency between IF and sound frequencies.

Figure 4.13. A ceramic filter for use at 6 Mhz (SFE 6.0 MB). Approximately twice actual size

At some stage in the video amplifier, the 6 MHz intercarrier signal is tapped off for processing in the sound circuits. We need to get rid of the vision signal, which at this point is many times greater than the wanted sound carrier. What is required is a sharply-tuned filter centred on 6 MHz, with a bandwidth sufficient to pass the sidebands of the sound signal – say 300 kHz. A *high-Q* tuned circuit can be used for this, but more commonly a ceramic filter will be found, as pictured in *Figure 4.13*. This is a tiny piezo device which has an electromechanical resonance (for TV IF versions) of 6 MHz, and simulates a very high-Q tuned circuit. After filtering by this, the intercarrier signal now undergoes several stages of amplitude limiting. This invariably takes place inside an IC, the limiting being necessary to strip off any amplitude modulation which may be present due to the vision signal or interference effects. This limiting is a common feature of FM

98

receivers, and nowhere is it more necessary than with the TV intercarrier system.

Referring again to *Figure 4.8*, the limiting takes place inside the sound IF chip, *IC302*, which also incorporates the sound detector. Many types of circuits have been used in the past for demodulating an FM signal, but in modern television practice, a *quadrature* detector is generally used. This is very similar to the synchronous detector described above for vision detection, but this time the applied reference is held one quarter of a cycle away from the incoming 6 MHz carrier, hence the term 'quadrature'. This means that when the incoming carrier is exactly 6 MHz (i.e., unmodulated), the bridge becomes conductive just at the moment when the carrier is passing through zero – hence no output. When the sound carrier is modulated, however, it deviates from 6 MHz, so the phase, or timing, between the reference and carrier signals changes. This phase change means that the carrier is at some level above or below zero during the bridge conduction pulses, so an output is produced which varies in polarity and amplitude with the deviation of the sound carrier; *Figure 4.14* shows the principle. Again, the reference signal is developed across a coil external to the chip, *L304*, and this is carefully tuned to achieve exact quadrature conditions in the demodulator, corresponding to linear detection and clear sound.

Having now recreated an AF signal (at rather greater length than in a radio receiver!), all that is necessary is to pass it through a volume control and tone controls if fitted, thence to a conventional power amplifier. In most receivers, IC techniques are used here, coupled to as large a loudspeaker as the TV cabinet will accommodate. Sometimes the audio response in the amplifier circuit is tailored to compensate for the shortcomings of the speaker system, mounted as it often is in the far-from-ideal enclosure of a plastic television cabinet shell. An increasing trend is towards a single chip to cater for intercarrier IF, detection, AF amplifier, and power output, so this device, coupled with one or two ceramic filters and a loudspeaker, comprises the entire sound section of the receiver. In some designs, the second ceramic filter

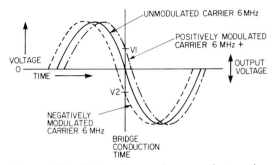

Figure 4.14. Timing diagram for a quadrature detector. As the frequency (or phase) of the modulated carrier swings up and down, the regularly-timed bridge-conduction pulse will catch it at a voltage level proportional to its deviation

takes the place of the critical quadrature coil, and this completely eliminates pre-set adjustments in the sound section of the receiver.

Remote-control systems

It is now time for us to back-track a little, and return to the 'front end' of the receiver. Many TV models offer cordless remote control so that programme changes and other adjustments can be made without touching the controls on the set itself. Where a receiver is fitted with Prestel or Teletext decoders, this feature is considered essential, and we are not aware of any such commercial designs which do not include this facility.

The link between the remote sender and the receiver is made at LF, either by ultrasonic sound or more usually by infra-red waves. In the latter case, the light waves are at low intensity and invisible to the human eye, being generated by an LED (light-emitting diode) and detected by a semiconductor form of photo-cell. The frequencies used for remote control usually lie between 30 KHz and 45 KHz. Sophisticated

remote-control systems are capable of 24 remote commands;
for example, 10 channel selections, six analogue functions
(i.e., volume up/down, brightness up/down, colour up/
down), six teletext commands, ideal picture (normalise), and
switch off. This is achieved by an IC in the remote sender,
powered by a small internal battery. When a remote command is given, the IC in the hand unit generates a *digitally-
encoded* signal which is sent over the infra-red link. On
receipt by the receiver, the command signal is picked up and
amplified, then passed to a decoder IC which has multiple
outputs.

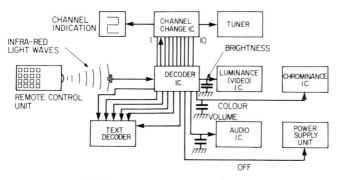

Figure 4.15. Block diagram of a typical full-specification remote-
control system

Let us assume that channel 2 is selected on the hand unit.
The encoder IC will modulate the transmitting LED with a
binary signal (i.e., a series of on-off pulses) which is interpreted by the decoding chip as 'go to channel 2'. The
appropriate output from the chip will be energised, and an
electronic switch (probably within another chip) will select
and indicate channel 2 and open the switch for whatever
channel was in use. If 'volume up' is the command, a
different binary code will be radiated, and the decoder chip
will begin to charge a capacitor in the receiver. The charge
will continue to rise while the command button is being
pressed, and the increasing DC voltage across the capacitor

101

is used to control the gain of the audio chip in the receiver. This is facilitated by the presence on the sound chip of a gain control pin, the amplification of the chip being proportional to applied voltage. In colour receivers, the *luminance* (video brightness) and *chrominance* (colour) processing chips have a similar facility, so that all three functions can be remotely controlled. *Figure 4.15* shows a typical remote-control system in block diagram form.

5

The receiver – timebases and power supplies

In the previous chapter, we have seen how the television signal is processed in the receiver, and how the vision and sound signals are recovered and applied to the picture tube and loudspeaker. Returning to the receiver block diagram (*Figure 5.1*), we will now explore the circuits that deflect the electron beam and provide the power to operate the set.

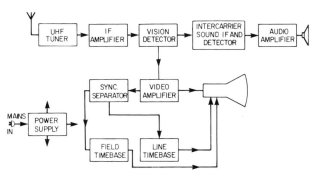

Figure 5.1.   Receiver block diagram

First, let us briefly summarise the points we have covered so far. The sound channel is operating and the picture, or video, signal is modulating the beam in the picture tube. Without any beam deflection, however, all we see is a brilliant pinpoint of light in the centre of the screen. We may notice that its average intensity changes, but the video signal

103

is geared to time, as we saw in Chapter 1, and as yet we have not introduced any time factor to the screen display. As we know, the timing of the video signal is governed by the regularly-repeated sync pulses, so we must return to these to begin a study of the timebase section of the receiver.

Synchronising separator

To make use of the sync pulses transmitted with the video signal, we must first separate them from the video signal itself, so we need a device to strip off the picture information, leaving only the sync pulses as such. This is not a difficult problem, because we can discriminate between the two on the basis of amplitude – for a standard (positive-going) composite video signal, the first 30 per cent (starting from zero, or ground potential) is sync information, and all

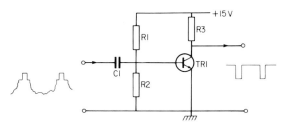

Figure 5.2. A simple one-transistor sync separator

above that is unwanted video signal. If we arrange an amplifier with a short tether, so to speak, it will delete the picture information by virtue of the fact that it will run into 'overload' at say 20 per cent of the full video input level. This means that it will swing from saturation to cutoff just on the sync pulses, and its output will consist of rectangular pulses corresponding to the inverse of the incoming sync. *Figure 5.2* shows the idea. Here we have a transistor amplifier which is biased by the resistor network *R1, R2* on its base so that it is just conducting. The composite video input, in negative-going form, is applied to its base via the coupling capacitor

104

C1. Because the applied sync pulses are positive-going, each one drives the transistor into *saturation* so that it switches hard on, reverting to cutoff for the 52 μs duration of the video signal in each line. Thus at *TR1* collector we have the sync signal, but it is in composite form – i.e., both line and field syncs are present. It will be recalled that the sync pulses instigate the retrace action of the timebases, and it is import-ant that the field timebase 'sees' only field sync pulses, and the line timebase only line sync pulses. Now we have to separate these, and certainly we cannot do this on an amplitude basis. Going back to *Figure 2.9*, we see that the line sync pulses have a duration of 4.7 μs, whereas the broad field pulses are 28 μs long. Here lies the key to distinguishing them.

Integrator and differentiator
Let us first consider the field pulses. We need a device which will produce an output only on a long-duration pulse, and the integrator fills the bill. *Figure 5.3 (a)* gives the bones of the circuit. The composite sync is fed in at the left side, and the short-duration line pulses do little to charge the capacitor *C*, being dissipated in the resistor *R*. When the field sync

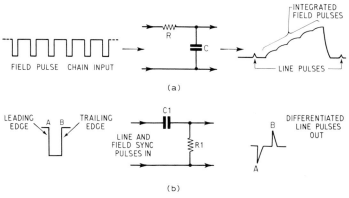

Figure 5.3. Sorting out the field and line sync pulses. (a) An integrator. (b) A differentiator

105

train comes along, however, not only are the pulses wider, but they are more crowded together, and so start to build up a charge on the capacitor. The resulting waveform is shown as a slightly jagged hump which, after cleaning-up, appears as a well-defined pulse coincident in time with the broadcast field sync group.

To extract the line pulses, we need a circuit with the opposite characteristic, known as a 'differentiator'. As *Figure 5.3 (b)* shows, it is merely an integrator turned on its side. The small series capacitor is charged each time a line sync pulse hits it, but the RC time constant is short enough to allow the capacitor to discharge quickly, say in $2\,\mu s$. Thus 'spikes' are produced, negative ones at the beginning of the line sync pulse, and positive ones $4.7\,\mu s$ later at the end of the pulse. These spikes form the line synchronising pulses, the un-wanted polarity spike being easily clipped off later by a diode. With such a short time-constant, the circuit is quite undisturbed in the presence of the field-sync train, merely producing its spikes on each pulse start and finish.

Now we have separate streams of sync pulses for line and field, we can apply them to the timebase generators to hold them in step with the scans in the camera tube. In practice, most or all of what we have been describing takes place in an IC, which also incorporates a noise-cancelling circuit to prevent interference and noise from upsetting synchro-nisation.

Field timebase

The function of the field timebase is to generate a sawtooth current waveform in the field-scan coils so that the beam will be deflected down the screen at an even rate and fly back to its starting point at the end of scan. Initially, the sawtooth is generated as a voltage ramp, then amplified in a power output stage to drive the scan coils. A representative field timebase is illustrated in *Figure 5.4*. The ramp capacitor is *C1*, and this charges through *R1* to give a linear rise of voltage at point *A*. Although the charging curve of a capacitor is exponential, a virtually linear ramp is obtained by using a very small initial portion of the charging curve (see *Figure 5.5*). When the charge on the capacitor has reached a certain level, the transistor switch is turned on and presents a virtual short-circuit to the capacitor, quickly discharging it. This corresponds to the field flyback, and the sawtooth so formed passes on to the power output stage. The turn-on current for the transistor switch is derived from an oscillator, often an *astable multivibrator* running at 50 Hz and synchronised by the field sync pulses. The free-running frequency of the

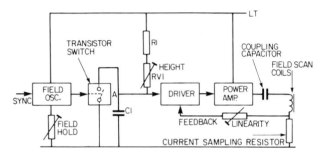

Figure 5.4. A field timebase. The ramp waveform is generated by the charge and discharge of *C1*

multivibrator is governed by the field hold control, and the ramp amplitude – and hence picture height – is governed by the charging current in *C1*. At low charging current, *C1* is not going to acquire much charge in the 20 ms field period, so the ramp amplitude will be low. If we reduce the value of the series resistor, charging current increasees, and in the same 20 ms the capacitor acquires a greater charge with a correspondingly higher ramp. Part of the charging resistor is made variable then, and this forms the height control, *RV1*. The sawtooth voltage thus formed is at high impedance and unsuitable for driving the field output stage directly. An intermediate amplifier (driver) is inserted and provides an impedance match between the ramp generator and the output stage.

The presence of the driver stage also affords an opportunity to correct the linearity of the vertical scan. In its passage through the driver and output stages and the scan coils themselves, the waveform undergoes distortions which cause the current ramp in the coils themselves to depart from a straight line. This of course tends to accelerate or decelerate the scanning spot on its journey down the screen (the same can happen with line scanning) and it has the effect of

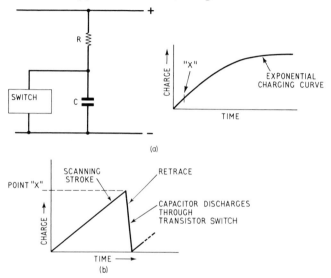

Figure 5.5. Forming the ramp. (a) The RC circuit produces a waveform that is linear up to point *X*, *C* then being discharged through the switch. (b) The resulting sawtooth waveform

stretching or cramping the picture in the area affected. If, for instance, the ramp tends to 'flatten off' as it approaches full amplitude, this will cause the downward-sweeping spot to decelerate towards the end of the field scan; the result will be that the horizontal scanning lines will be traced out closer together than they should be, and the bottom of the picture will be 'squeezed up' or cramped. To ensure a linear current ramp in the scan coils, feedback is introduced between the

108

output and driver stages. Often, the feedback waveform itself is developed in a small sampling resistor in series with the scan coils, so that any change in their performance with temperature is compensated for. The feedback is adjustable by means of one or two pre-set pots, which are adjusted for a linear scan.

The field output stage is a power amplifier with the scan coils as its load. Whether the field timebase is in discrete or IC form, almost invariably the *push-pull* technique is used, whereby two transistors are connected in series across the power supply, with the load connected to the centre point. As the lower transistor turns off, the upper turns on so that the power fed to the load is varied as a function of the transistors' input waveform. The idea is similar to that used in audio output circuits, and over the years many different configurations have been used to achieve push-pull operation. It is important that the transition from one of the output pair to the other is smooth (i.e., no *crossover distortion*). If this occurred, the spot would momentarily slow down at screen centre, and a cramping effect would take place there.

There is one other consideration in a field output stage, and this concerns flyback. As we shall see in the line timebase, a sudden change of current in an inductor (in this case, the field-scan coils) gives rise to a large voltage spike across it, and the field output stage must be able to accommodate this. If the flyback is hampered, it will take longer than its allotted 350 μs or so, and the start of picture information for the new field will catch the scanning spot on its way back up the screen. The incoming video information will now be superimposed on the existing picture, usually towards the top; this is described, from its appearance, as *foldover*, and the effect is shown in *Figure 7.3 (b)*.

We cannot connect the scan coils directly to the midpoint of the output transistors, however, as the DC then flowing would upset amplifier operation and probably deflect the picture off the screen altogether. The coupling is made via a large capacitor, large enough to have a low *reactance* at 50 Hz, so as to form a DC block but pass the sawtooth current unscathed into the scan coils.

The field timebase, being a relatively low-power stage, fell early to the invading march of IC technology. First, chips intended for audio applications were pressed into service for small-screen receivers, but for some years purpose-designed ICs have been available which embody the oscillator and driver stages as well as the power amplifier. Such a device appears in *Figure 5.6.* All the features of the discrete timebase are there. The incoming sync signal undergoes a double integration in *R501/C501/R502/C502* before being applied to

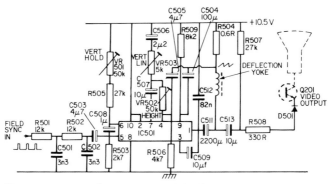

Figure 5.6. Complete commercial field timebase using one IC (NEC Semiconductors and Sanyo Ltd)

IC pin 5. The time-constant of the field oscillator is governed by the components connected to pin 6. These are *C508* and *R505/VR501*, the latter forming the vertical-hold control. The ramp charging capacitor is *C505* with amplitude control by height pot *VR502*. The field output appears on pin *1*, coupled into the scan yoke by *C511*. A small-value resistor is inserted in the scan-coil return lead (to the HT line, not ground this time) for current sampling purposes, the resulting voltage sawtooth going back into the chip on pin 9 to provide negative feedback for scan linearisation. One field linearity control is provided, *VR503*.

In the previous chapter, we mentioned the necessity of blanking the scanning spot in the picture tube during flyback. Here it is accomplished by a feed from the 'hot' side of the scan yoke via *R508* and clipping diode *D501* to the emitter of the video output transistor. The positive-going flyback pulse produced by the inductive scan coils momentarily raises the emitter potential of the video transistor and causes it to cut off for the duration of flyback; the pulse is amplified in the transistor, and a positive pulse is passed to the cathode of the picture tube. The scanning spot is thus extinguished until flyback is complete.

In the design of the field timebase, there are some later developments. For large-screen receivers, a class D field timebase chip is in use in some models. It works on the 'switch-mode' principle and is designed to overcome the power-dissipation problems inherent in conventional (class B) push-pull systems. The idea is that the output stage works as a switch rather than as a linear amplifier, the switching rate being very much greater than the 50 Hz field repetition rate. At the beginning of scan, the switch spends most of its time off, so energy pulses at the output terminal are relatively few and far between. As the scan proceeds, the ratio of on-time to off-time is gradually increased, so by the end of the field scan period (bottom of picture), the *mark-space ratio* has increased to the point where the switch is on for say 95 per cent of the time, and a lot of energy is being delivered to the scan coils. This system has the advantage that the switch in the chip is either on or off at any one time, and in neither state does it dissipate much energy. As a result, the chip runs cooler and more efficiently than a class B amplifier.

Another innovation made possible by IC technology is the *indirect field sync* system. This works in conjunction with a 31.25 kHz (twice line speed) oscillator whose output is divided by 625 to provide a field sync pulse, and by two to furnish a line timing pulse. This countdown process is just the same as that used in some sync generator circuits (see, for instance, *Figure 2.10*), and while the derived 50 Hz pulse cannot be used directly to synchronise the field timebase because it has no particular timing reference to the picture, it is used to check the accuracy of the received field sync pulse and to correct the field synchronisation in the presence of interference and sync-pulse distortion.

111

Line timebase

The line timebase is more complex than its field counterpart for several reasons. A more sophisticated synchronising circuit is used, and the relatively high repetition rate of 15.625 kHz means that the scan waveform is more difficult to produce than the much slower field scan. In addition to this, the energy produced by the line output stage is harnessed to provide other services than beam deflection. To recapitulate the requirements of a line timebase: its basic function is to deflect the scanning spot in the picture tube from left to right in synchronism with the beam in the camera. So we need an oscillator to generate the line scan, a means of locking it to the incoming sync pulses, and a power output stage to drive the scan coils and generate voltages for use elsewhere in the receiver. Let us look at each stage in turn.

Line oscillator

As we shall see when we come to study the line output stage, the sawtooth shape of the line-scan current waveform in the scan coils is created by the inductance of the line output transformer and scan yoke itself – not by any special characteristics of the waveform coming from the line-scan generator, or oscillator. Over the years, many types of oscillator have been utilised in line generating circuits – multivibrators, blocking oscillators, and often sinewave oscillators using an LC tuned circuit to generate the basic frequency. The fact is that any of these are capable of generating the timing pulses, which is all that is needed by the line output stage, the active part of which is very little more than a high-speed on-off switch.

An LC oscillator is shown in *Figure 5.7*. The tuned circuit is formed by the coil *L35* and the split capacitor *C17a–b*, which is tuned to be resonant at line scanning frequency, 15.625 kHz. A portion of the energy in the tuned circuit is passed into the base of *T27*, and appears amplified and inverted at its collector. In a periodic circuit like this, the signal inversion has the same effect as delaying the signal by half a cycle – i.e.,

112

a 180° phase change. Energy from the collector is fed back to the tuned circuit via *R26*, and its phase is such as to reinforce the energy in the LC tuned circuit, and maintain oscillation. Thus a *sinusoidal* waveform is set up across the inductor *L35*, and the base bias on *T27* is set by *R27* to a level such that the transistor operates virtually in *class C* – that is, it is driven from cutoff to saturation by the waveform appearing at its base. Hence the waveform appearing at its collector consists of rectangular pulses at line-scan rate. As in the field circuit,

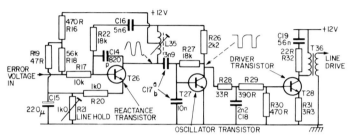

Figure 5.7. Line generator and reactance stages for a monochrome TV (Philips)

the line oscillator is not capable of driving the power output stage itself, and a buffer/amplifier circuit is inserted between them, here consisting of the line driver transistor *T28* and matching transformer *T36*. The secondary winding of *T36* provides a low-impedance drive to the line output stage, which we shall meet shortly.

Line synchronisation

In early designs of TV receiver, the line sync pulses were directly applied to the oscillator to hold it in step, just as in the field timebase. In the latter, synchronisation is satisfactory because the broadcast field-sync pulses are broader than those used for line sync, and the field-triggering pulse itself consists of the integrated product of several such pulses. Thus the field sync signal has some 'meat' in it and is quite robust. In line synchronisation, however, things are a little

113

different. Because of the negative vision modulation system used by broadcasters, the sync pulses are the most positive part of the broadcast signal and so are very vulnerable to interference. Impulsive interference often consists of sharp spikes, and to the receiver circuits these look very similar to line-sync pulses. An interference spike arriving just before the due time for a sync pulse would cause the timebase to trigger early, so the scanning line thus initiated would be displaced sideways. In an interference situation, the picture would appear ragged; in poor-reception areas and 'high-noise' environments, this raggedness would be barely tolerable.

In mechanical engineering, the problem of intermittent mechanical motion is solved with a flywheel, heavy enough to iron out the variations in input energy by means of the kinetic energy stored in it. To iron out time variations in line-sync pulses, we can use an electrical equivalent of a flywheel – simply a storage capacitor to average out the effect of a large number of incoming line-sync pulses, so that mistiming or distortion of individual pulses will have no effect on the display. It is quite simply done: the frequency of the line oscillator in the set is compared with the frequency of the received sync pulses. If any difference is detected, an error voltage is produced and used to modify the charge on a capacitor. The charge on this capacitor indirectly affects the oscillator frequency, pulling it back into step whenever it wanders off. Because of the relatively long time-constant of the capacitor (heavy flywheel effect), random noise and short-term sync-pulse distortion will have little or no effect on line synchronisation.

How is this done in practice? *Figure 5.8* shows a skeleton circuit of one of the many variants of the flywheel sync circuit. The incoming sync pulses are applied to a transformer whose secondary winding produces equal outputs of opposite polarity. These are used to simultaneously switch on a pair of diodes connected in series. Applied to the junction of the diodes is a sample sawtooth waveform, derived by integration of a line pulse from the line output stage. In the absence of sync pulses, this waveform gets nowhere, as neither diode can conduct. On the arrival of a sync pulse, however, both diodes conduct; the instantaneous level of the sawtooth waveform at that moment is sampled, and apears at the outer ends of the diodes, and via R2 and R3 at the flywheel capacitor C3. Thus the storage capacitor C3 receives a charge during each sync pulse via the conducting diodes, and the amount of charge it receives depends on the voltage level of the sawtooth at that moment. If the sawtooth waveform happened to be passing through zero at the moment the diode gates opened, the charge on the capacitor would be zero. What happens if the oscillator speeds up? The sync pulse now appears to arrive too late, and sampling takes place further up the ramp, so that a

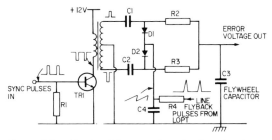

Figure 5.8. Flywheel line synchronisation circuit

positive pulse is conveyed to C3. Conversely, if the oscillator slows down, sampling takes place on the negative portion of the ramp, and C3 is charged in a negative direction. Thus the charge on C3 and its polarity depend on the magnitude and direction of the frequency error in the line oscillator; if the oscillator wanders a little down in frequency, a small negative charge will appear. If, alternatively, the oscillator is running at a frequency higher than 15.625 kHz, a large positive charge will be acquired by C3. This process as described is known as *phase* discrimination, as it compares the phase, or timing, of the local and incoming line pulses. The circuit also works as a frequency discriminator, producing the correct sense of

115

error voltage when the line oscillator is running at a frequency above or below that of the incoming sync pulses. All that remains now is to apply this error voltage to the line oscillator so as to control its frequency, and flywheel synchronisation has been achieved.

Reactance
It is a basic characteristic of a capacitor that the current flowing in it leads the voltage across it by one quarter of a cycle (i.e., 90°). If we can arrange a circuit in which this condition is present, an oscillator can be 'fooled' into seeing it as a capacitor. In *Figure 5.7*, the transistor *T26* is set up by *C14, C16,* and *R22* so that its base current leads its collector voltage by 90°. Thus it appears like a capacitor, the effective capacitance of which varies according to the base bias applied. This voltage-controlled capacitance effect is called 'reactance', and the error voltage from the flywheel capacitor is applied to the base of the transistor *T26* via filtering circuits. The collector of the reactance transistor is connected to the tuned circuit, so that the oscillator (*L35/C17*) is capacitively loaded to an extend depending on the flywheel error voltage. Initially, the reactance of *T26* is set up by the line-hold control *R21*, and the coarse hold control is *L35*. The *phase lock loop* is now complete, and we have a stable line lock, immune from most interference.

ICs for line generation
Although we have gone through the processes of line synchronisation and generation at some length to explain the principles involved, all commercial colour receivers (and many monochrome ones) in production use a single chip for these purposes. The principles of operation are the same, although the line oscillator does not use an LC tuned circuit, depending rather on an external precision capacitor to set line speed, in conjunction with a resistive line-hold control pot. For small-screen applications, the line driver stage is also built into the chip; a commercial circuit for use with a *Darlington* line output transistor is shown in *Figure 5.9*. The TDA1180 chip contains (in addition to the sync separator,

116

flywheel circuit, line generator, and line driver) a noise-cancelling circuit for both line and field sync, further enhancing sync performance in poor conditions. Composite video is taken in at pin *8*, the line sync pulses produced therefrom being compared with line flyback pulses passed into pin *6* and integrated by *C59*. The flywheel filter capacitor is *C57*, and the oscillator frequency is set by *C58* and *R79/RV8*. The output from the chip directly drives the line output transistor via *C62*, and the whole operates from an 11 V line. In this particular set, the field sync pulses coming out of the chip on pin *10* directly feed the field generator chip.

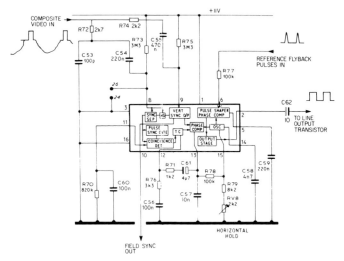

Figure 5.9. Line generator using an IC. The chip contains all the circuitry required from video in (for sync) to line drive out (Thorn Consumer Electronics Ltd)

There is one section of the receiver which nobody has managed to 'chip' so far – the line output stage. Let us follow our carefully synchronised line pulses into this department and see how they are made to deflect the scanning spot and build up the picture.

Line deflection

It is a characteristic of an inductor that when a constant DC voltage is applied to it, the current flowing in it rises linearly from zero. Since this is just the waveform we require for line deflection in our TV receiver, it may be thought that all we require is a switch to apply DC to our scan coils once each line. Basically, this is exactly what does take place, but with a little refinement for the sake of efficiency. A much simplified circuit of a line output stage is shown in *Figure 5.10*. The line output transistor may be regarded as no more than a switch, capable of grounding the bottom of the line output transformer (LOPT) primary whenever it is switched on by the line

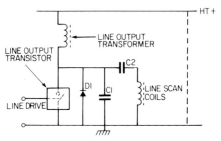

Figure 5.10. Basic line output stage

drive waveform. When the line drive waveform turns the switch on, the primary of the line output transformer is connected directly to the DC supply voltage. Thus a linear build up of current takes place in the LOPT and in the scan coils. This has the effect of deflecting the beam in the picture tube from the screen centre to the right-hand side, and when the spot has reached this point, the transistor switch is suddenly turned off by the cessation of the line drive pulse. The tuning capacitor $C1$ is now rapidly charged up by the energy present in the magnetic field of the LOPT, and current flows back out of the LOPT into $C1$. This current reversal in the LOPT and scan coils causes the scanning spot to fly rapidly back to screen centre, and it keeps going beyond this point as the LC circuit formed by the LOPT and $C1$ complete a half-cycle of oscillation, with energy now flowing back from

118

C1 into the LOPT. This to-and-fro exchange of energy between the L and C elements is what normally takes place in any tuned circuit, and the one half-cycle is sufficient to take the scanning spot right back to the left-hand edge of the screen. If left to its own devices, the LOPT and tuning capacitor would continue to *ring*, but the process is interrupted at the end of the first half-cycle when the top plate of *C1* tries to go negative. At this point, the efficiency diode *D1* becomes forward-biased and turns on, damping out any further oscillation and clamping the LOPT across the DC line again. The magnetic field present in the LOPT now decays linearly to zero, as does the current in the scan coils. This brings the spot back to screen centre. We have now completed one horizontal scanning cycle and traced out a single television line. There is no energy left in the LOPT or tuning capacitor. At this point (in practice a little earlier), the line output transistor (switch) is again turned on, and away we go to trace out the next line; the cycle repeats itself 15 625 times every second, and line scanning is thus achieved. the current in the scan coils is the same as that in the LOPT; in fact, the two components may be regarded as being virtually in parallel, and the inductance of the scan yoke reinforces that of the LOPT.

We have not yet mentioned the coupling capacitor *C2*. As in the field-scan circuit, we cannot allow any DC in the scan coils; the permanent deflection field thus set up would deflect the beam to one side. *C2* is a convenient DC block, and in wide-angle sets performs another useful function known as 'S-correction'.

S-correction
If the screen of a picture tube were spherical rather than flat, the electron beam would have the same distance to travel to the screen regardless of the amount of deflection it suffered. Because the screen is flat in practice, the beam's path-length varies with deflection angle, being short at screen centre and longest at maximum deflection (at the corners). If we use a true sawtooth deflecting current in these circumstances, the picture will be stretched at the edges and cramped in the

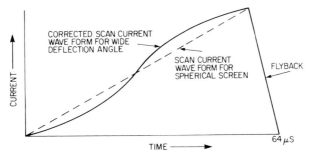

CORRECTED SCAN CURRENT
WAVE FORM FOR WIDE
DEFLECTION ANGLE

SCAN CURRENT
WAVE FORM FOR
SPHERICAL SCREEN

FLYBACK

CURRENT

TIME

64 μS

Figure 5.11. The modified shape of the scanning current for
S-correction in wide-angle picture tubes

middle. If we can arrange to slow down the rate of change of
the scan waveform at the beginning and end of scan, we can
compensate for this, and the value of the coupling capacitor
C2 is chosen to give the curve required (see *Figure 5.11*). The
resultant waveform has a characteristic S-shape – hence the
term 'S-correction'.

Auxilliary supplies

The voltage across an inductor is proportional to the rate of
change of the current flowing in it. During the fast current
reversal which forms the flyback stroke, the rate of change of
current in the LOPT is greatest, and this gives rise to a large
voltage pulse across the LOPT during line flyback. This
voltage pulse is convenient for driving other sections of the
receiver, and almost any required voltage can be obtained by
rectifying flyback pulses from the LOPT. In Chapter 1, we saw
that the picture tube needs various operating voltages on its
electrodes, principally a very high accelerating voltage on its
final anode and a few hundred volts on its first anode. In the
practical circuit of *Figure 5.12*, we can see how these are
derived. This circuit is used in a small-screen monochrome
receiver with a 35 cm tube whose EHT requirement is 11 kV or
so. This is obtained by stepping up the flyback pulse amplitude in a secondary winding and rectifying the result in *D10*.

120

For large-screen and colour receivers, a *voltage multiplier* often takes the place of *D10*; this is a rectifier/capacitor combination which doubles or trebles the secondary voltage to provide EHT. Other voltages are produced in like fashion. *D9* and *C90* provide a 130 V line to operate the video stage and picture tube first anode, while 14 V and 26 V lines are furnished by *D8/C85* and *D7/C86* respectively. In many sets, the picture-tube heaters are also powered from a winding on the LOPT.

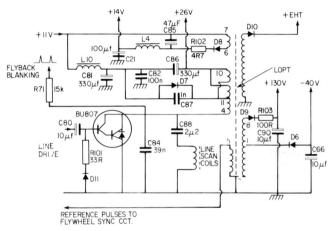

Figure 5.12. a practical line output stage (Rank-Bush-Murphy Ltd)

The circuit in *Figure 5.12* is one of the simplest in commercial use; in more sophisticated sets, the line output stage is more complex, but the basic principles remain the same. Very often, the efficiency diode is nowhere to be seen – its function is taken over by the line output transistor, either by means of a built-in diode, as in the present circuit, or, where a driver transformer is used, by the diode formed by the *c–b* junction of the transistor itself. In this case, the clamp path to earth is via the low-impedance secondary winding of the line drive transformer.

The other elements of our basic circuit are easily found in *Figure 5.12*. The tuning capacitor is *C84*, S-correction capacitor *C88*, and the LOPT primary winding is between pins *4* and *11*. In our study of flywheel line sync, we saw that a reference pulse was required by the discriminator circuit for use in checking and correcting the frequency and phase of the line oscillator. In *Figure 5.12* this comes from pin *5* of the LOPT. Flyback blanking is necessary for the line scan as well as field scan; the flyback pulse at *TR9* collector is taken off through *R71* and applied to a point in the video amplifier to achieve beam blanking during line flyback.

Power supplies

As we have discovered, the line output transformer is a handy source of power for other sections of the receiver. In many sets, large and small, the primary power source, be it mains or battery, is used to feed the line output stage only, all other supplies being derived from the latter. In other designs, the main HT line feeds the field timebase and signal processing circuits directly. In all cases, the set's HT line must be stabilised against fluctuations in supply voltage and current demand by the set itself. Picture size and EHT are proportional to HT voltage, and we cannot have these fluctuating. We will examine two forms of HT stabiliser, one at each end of the television 'social scale'.

Figure 5.13 illustrates a series stabiliser as used in many portable monochrome sets. These are designed to run from mains via a transformer and rectifier, or a 12 V DC source such as a car battery. The input voltage can vary between 12 V and 16 V, and the heart of the circuit is the series regulator *TR5*, which is made to absorb the difference between the incoming voltage and the wanted HT line voltage of 11 V. The transistors *TR6* and *TR7* are arranged to compare the 11 V stabiliser output with a fixed reference voltage. Any deviation from the correct level is sensed and turned into an error current which is used to turn *TR5* up or down to compensate.

The 11 V line is developed across a potentiometer network *R55/RV4/R56*, so that the base voltage and current in *TR7* is proportional to HT voltage. Let us assume that a bright picture has increased the beam current in the picture tube, resulting in an increased current demand by the line output stage. This would tend to load down and lower the HT line, so that it would start to fall, reducing base current in *TR7*. This would result in less current through its emitter resistor *R54*, and the voltage across it would fall. The emitter of *TR6* is connected to this point, and because its base is connected to a fixed reference voltage (derived from 5.6 V zener diode

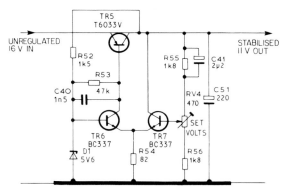

Figure 5.13. Series regulator for small-screen portable TV (Thorn Consumer Electronics Ltd)

D1), the current in *TR6* increases. Now *TR6* collector current flows through the base of the PNP regulator transistor *TR5*. With an increased base current, *TR5* turns harder on to compensate for the increased current demand, and the line voltage is pulled back to 11 V. Thus is the stabilisation loop formed. The process works in reverse, of course, and also compensates for variations of input voltage, so that the HT line voltage depends only on the setting of the 'set volts' control *RV4*.

This stabiliser system is fine for portable TV use, but it is wasteful of energy in that the series element *TR5* is in effect no more than a variable series resistor and dissipates unwanted energy as heat. In many mains-operated receivers, a more efficient system is used to derive a stable operating voltage, on the 'chopper' principle. We briefly met this in the field timebase section, and again the idea is that the regulating element is in the form of an electronic switch, alternating between on and off. The ratio of on-time to off-time is varied in sympathy with the current demand of the receiver, so that the energy taken from the mains, when averaged out over several switch cycles, is just that required by the set, without wasteful dissipation of heat energy in the power-supply unit itself. The HT operating voltage for a TV set is dictated basically by the requirements of the picture tube and its attendant line output stage, and current designs need an HT rail voltage between 130 V and 200 V. Let us see how this is obtained in a *switch-mode* power supply unit (PSU).

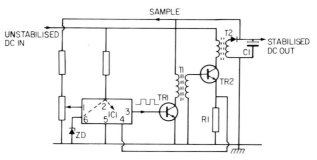

Figure 5.14. Basics of a switch-mode power supply using a purpose-designed IC

The diagram in *Figure 5.14* shows a skeleton circuit of the arrangement. The purpose-designed chip, *IC1*, generates a square wave which is amplified in the chopper drive transistor *TR1* and used to turn on and off the chopper switch itself, *TR2*. Thus the primary winding of the chopper transformer is periodically connected to the mains-derived and unstabilised DC input. This of course sets up a magnetic field in the core of *T2*, and one or more secondary windings are excited. The pulses appearing across these are rectified and used to provide the DC operating voltages for the set, in a similar way

124

to the flyback rectification we met in the line timebase department. This is fine, but we have not stabilised the output voltage yet.

A sample of the rectified output voltage is 'potted down' and fed back into the chip on pin 1. This is compared with the local reference voltage derived from the zener diode on pin 6. Assuming that the HT line voltage developed on C1 started to rise, the error signal within the chip would take the form of a narrowing of the drive pulse coming out of pin 3. This would cause TR2 to dwell longer in the off state, and the net energy fed into the chopper transformer, and hence the receiver circuits, would decrease, restoring the output voltage to normal. The feed to pin 4 of the chip is representative of the several forms of overload protection built in to such a circuit. In this case, the current in the chopper transistor is monitored by R1, and in the event of an abnormally high current, the chip is signalled via pin 4 to reduce the mark-space ratio to minimum or shut down the PSU completely. Such an overload may be caused by a short circuit or malfunction in the receiver circuits fed from the PSU.

This chopper regulation system has much to commend it; very little energy is wasted in the PSU itself, and because the whole thing works at high frequency (usually the 15.625 Hz line rate), the chopper transformer can be a small and relatively cheap, ferrite-cored device. Another advantage is that the receiver metalwork can be isolated from the mains if the chopper transformer primary and secondary voltages are insulated from each other. There are many variations on the chopper or switch-mode theme in use by set makers. Because of the similarity between switch-mode PSUs and line output stages (compare *Figures 5.12* and *5.14*), it is possible to integrate the two functions, and this has been done in commercial designs of monochrome and colour receivers.

Monochrome TV

We have now looked at all aspects of the monochrome receiver in basic form, and seen how the signal is obtained

125

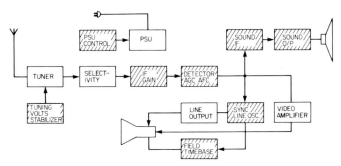

Figure 5.15. 'Chip' areas of a TV receiver. The hatched blocks show the stages which are currently amenable to IC technology

and processed to modulate the scanning spot. We have studied the generation of the scanning fields and their synchronisation, and seen how the operating voltages for the set are derived. *Figure 5.15* shows the areas of the receiver which are amenable to IC technology, and how little is left in discrete form.

In the next chapters, we shall see how the principles and techniques of basic television are expanded into the areas of data transmission, video recording, and colour TV.

126

6

Colour television

We have seen that monochrome television relies on a fast-moving scanning spot which is able to change swiftly in brightness between black and peak white to shade-in, as it were, in black and white a picture on the screen of the picture tube which is a replica of the image of the televised scene on the photosensitive screen of the camera tube.

Colour television works in virtually the same manner, but the effective scanning spot on the colour display tube is able not only to change to peak white from black but also in hue and colour saturation over almost all the colours of the rainbow. Colour television uses the line and field scanning of monochrome television, with the line-repetition frequency fixed at 15 625 Hz to form a 625-line picture with a field frequency of 50 Hz (the British standard). A colour TV receiver is a monochrome set with the addition of a decoder and with a special picture tube. Every feature of the monochrome set is present, and except for the video processing after the detector, all the circuits and processes are the same. In a colour set, more power is required by the picture tube, so the power supply and timebases are 'beefed up' to suit, though they do just the same job.

Because 625-line transmissions carry monochrome programmes as well as colour ones, colour sets are designed to respond automatically to the type of transmission picked up by the aerial, so that a monochrome transmission gives a picture in black and white and a colour transmission gives a colour picture without the set having to be switched in any way. This is called *receiver compatibility*. Compatibility is also built into the transmission itself, so that a colour transmission picked up on a black-and-white recever will give a good monochrome picture ('reversed compatibility').

Receiver compatibility and reversed compatibility demand that the signals carrying the colour information of the televised scene be integrated within the existing 625-standard monochrome signals in such a manner that they are rejected by a monochrome set and processed only by a colour one. The fundamental requirement is that the colour transmitter (and studio equipment) must generate ordinary monochrome television signals in addition to colour ones.

In colour-television parlance, we refer to the monochrome components of a colour signal as *luminance signal*, which is symbolised by the capital letter Y, and the colour components as they eventually pass through the colour set as *chrominance signals* or *colour-difference signals*, for the reason which will be obvious later.

At the television camera, the Y, or luminance, signal is the same as the brightness monochrome signal of a black-and-white television system.

In some colour cameras, there are three camera tubes, instead of the single tube in a monochrome camera. These are arranged to respond separately to the red, green, and blue light in the scene on which the camera is focused. In other words, the scene is analysed in terms of the three primary colours – red, green, and blue – by the three tubes. This analysis is achieved simply by colour filters or reflecting systems.

Almost all colours in nature are composed of mixtures of the three primary colours. When an artist sees the colour of his paints, his eyes are responding to coloured light reflected by them. A paint of particular hue ('hue', incidentally, is just another word for colour) *absorbs* all colours except that which it is reflecting. A mixture of paints absorbs all the colours except those which they are reflecting collectively. Pigments are thus given colour by a subtractive light-mixing process.

128

In colour television, however, we are dealing with coloured lights. These are mixed by an additive process: the colours present in the scene and radiated from lights on the screen of the display tube are added *directly* by the eye.

When correctly proportioned amounts of red, green, and blue *light* are mixed, we get white. Conversely, white light is composed of the three primary lights – red, green, and blue. It follows that pretty well all the colours of the rainbow can be created from the three primary hues by mixing their lights in various proportions.

Moreover, all shades between black (no light) and peak white – in other words, the various greys – can be obtained if the red, green, and blue lights *in the proportions required to give white light* are *together* varied in strength. That is, simply by increasing the intensity of the lights (from zero) while ensuring that their proportions in the mix remain constant as for white light, we obtain all the possible shades of grey. It is on this basis that compatible colour television works.

The camera produces signals corresponding to the amounts of red, green, and blue light in the scene, while the picture tube at the receiver produces an effective scanning spot made up of three red, green, and blue lights. Thus this spot can be made to change from black to peak white and also in colour to display almost any hue.

The colour information given out by the camera is encoded upon the ordinary monochrome information, also given by the camera as we shall see, while at the colour receiver a decoding circuit extracts the colour information and translates it into signal voltages which regulate the overall colour of the effective scanning spot so that this matches at any instant in the scanning process the colour of the picture-element at the camera tubes. The monochrome information (that is, the luminance or Y information), which is also fed to the colour-display tube, controls the correctly proportioned intensity of the three lights making up the effective scanning spot.

The Y signal thus operates the three lights proportioned for white on a monochrome transmission in exactly the same way as it operates the one light (scanning spot) of a

129

monochrome set. On colour transmissions, however, the Y signal regulates the luminance of the colour display, and further, colouring, information is added.

There is not sufficient room in this rather important chapter to delve deeply into the complications of pure optics and colorimetry, and it is best to concentrate on the practical aspects of colour television proper. However, the reader should know a little about light and colours.

The light spectrum

Light is an electromagnetic radiation, just as radio waves are, and our eyes are receivers of this radiation. The wavelength of light, though, is very much shorter than radio or television signals, and it does not occupy just one wavelength but

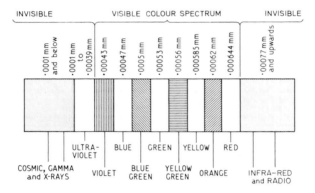

Figure 6.1. The light spectrum, showing wavelengths in millimetres

spreads out into a small spectrum, called the 'light spectrum', corresponding to all the colours of the rainbow. *Figure 6.1* shows the light spectrum sandwiched between small spectrums of invisible radiations, such as cosmic, gamma-, and X-rays at the high-wavelength end, and with radio radiations at the low-wavelength end.

130

Our eyes take in all the colours or wavelengths together when we see white light. If, however, the radiation is concentrated at some point in the spectrum, we see colour corresponding to that radiation.

Indeed, our eyes analyse light in a manner similar to that in which colour-television engineers endeavour to get their colour cameras to analyse it. The eyes contain the two types of cell, called 'rods' and 'cones'. The rods provide the brightness or luminance sensations of visions, while the

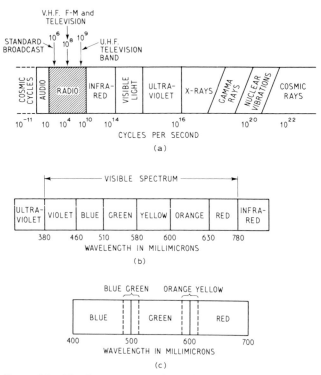

Figure 6.2. The frequencies and wavelengths of radiations. (a) The electromagnetic spectrum. (b) The visible spectrum, with individual colours. (c) The three primary colours

131

cones provide the colour sensations. There are cones responsive to the three additive primaries, red, green, and blue, as there are three camera tubes corresponding to these colours in a colour camera. Some colour cameras, as we shall see, have a fourth tube dealing essentially with the Y information, which is analogous to the rods of the eye. Three-tube cameras, however, add the red, green, and blue processed information to give luminance.

Figure 6.2 shows *(a)* the complete electromagnetic spectrum, taking in colour, *(b)* the colour spectrum in greater detail, and *(c)* the three primary colours of television in spectrum. These diagrams indicate the frequencies and wavelengths of radiations.

If we hold a coloured transparent screen or filter in front of our eyes, the scene at which we are looking will be analysed in terms of that colour. A red screen will let through only the red light from the scene. This is a red-pass filter. Green- and blue-pass filters will let through only green and blue light. This is basically how a scene is colour-analysed in a colour-television camera.

Basic colour system

A good way of looking at the basis of a colour-television system is illustrated in *Figure 6.3*. Here we have the three camera tubes with red, green, and blue filters in front of them. The tube with the red filter gives a red video signal, that with the green filter a green video signal, and that with a blue filter a blue video signal. If these video signals are separately applied to three picture tubes with corresponding colour filters in front of them and arranged in an optical system so that the three pictures are exactly superimposed upon each other, a composite picture in full colour will result.

While this system may be suitable for closed-circuit television, it is not any good for off-the-air domestic colour television. The main reason is that it would call for three separate transmitters, one each for red, green, and blue video signals, *plus* of course, a sound transmitter. Similarly, each set would have to have a total of four receivers and three colour picture tubes.

Not only would this system be wasteful of radio-space (if this were available to waste, which it is not) but it would fail to fit in with the requirements of compatibility, for a black-and-white set would receive at any time only one of the colour channels, depending on its tuning.

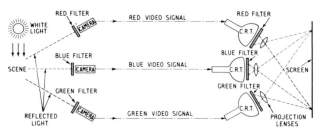

Figure 6.3. An elementary colour television system, in which the televised scene is broken down into the three primary colours

We shall see later how the colour information is neatly sandwiched between the components of monochrome information in a colour transmission, thereby allowing it to be carried in a single television channel along with the ordinary monochrome information.

However, before we get on to the encoding and decoding of colour information, let us have a look at a practical colour-television camera system and display tube, the latter being built in the form of a single tube, as distinct from the three tubes shown in *Figure 6.3.*

Inside a colour camera

A practical colour-television camera is shown in *Figure 6.4.* There are still three camera tubes, one for each primary colour, but only a single lens-system and housing. The light from the scene passes through the lens in the normal way, and this focuses the image of the scene simultaneously on to

133

the photosensitive surfaces of the tubes, each tube dealing only with its own corresponding colour, as in *Figure 6.3*.

Inside the camera housing are optical devices for analysing the light and directing it in colour isolation to the three camera tubes. These are not the filters of *Figure 6.3*, but special mirrors, called *dichroic mirrors*. They have the remarkable property of reflecting light of one primary colour while letting through light of the other two primary colours.

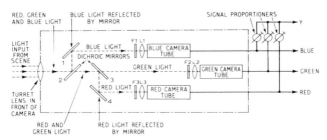

Figure 6.4. The optics of a practical colour television camera

Dichroic mirror *2*, for instance, lets through red and green light from the scene and reflects blue light, while dichroic mirror *3* lets through the green of the red and green applied to it and reflects the red. Separate, ordinary mirrors, *1* and *4*, reflect the blue and red lights respectively on a line corresponding to the blue and red camera tubes. These are ordinary tubes, but their light input is of one colour only, so we can call them after the colour that they handle. Similarly, we can call the signal that each one delivers after the corresponding colour. The signals cannot be coloured, of course, but one talks of red, green, and blue video signals, referring to the tube from which they were derived.

F1, *F*, and *F3* are final tailoring filters to balance the spectral responses of the tubes; *L1*, *L2*, and *L3* are final focusing lenses whose adjustment, with the adjustment of the mirrors, allows faithful registration of the three colour-isolated images on the three tubes. Correct registration is essential to avoid colour-fringing symptoms on the received picture.

134

Tubes used in colour cameras are the photoconductive vidicon type (see page 26). Each tube in a colour camera operates in the way explained in Chapter 2, and their electron beams are deflected in perfect synchronism.

Some colour cameras incorporate a fourth tube which receives all the components of light of the scene, as does the single tube of a monochrome camera. Such a camera delivers a monochrome (or Y) signal in addition to the red, green, and blue signals.

Deriving the Y signal
Anyway, there is no problem in delivering a Y signal, since this can be obtained by combining the red, green, and blue video signals in the correct proportions. Each tube system is initially set up to deliver the same signal output when the camera has a white light input, but these signals are subsequently adjusted at the output of the camera proper so that the red signal is 30 per cent, the green signal 59 per cent and the blue signal 11 per cent. These proportions are very important and should be remembered, as should the fact that the largest percentage of signal is taken from the green tube and the smallest from the blue tube.

When these proportions are added together, we have a signal which is identical to that delivered by the single tube of a monochrome camera. This is the Y signal. The proportions are selected by the proportioners at the camera output shown in *Figure 6.4*. The proportions add up to 100 per cent, which means that the Y output will be the same as the combined red, green, and blue outputs when the camera has a white input.

The shadowmask colour tube
We will now have a look at the colour-display device used in domestic colour sets. This looks just like any ordinary monochrome picture tube from the outside, but it has a very different scan yoke and is somewhat heavier than its monochrome counterpart. It comes in various sizes from 13 cm to 68 cm, though the most popular sizes in the UK are 51 cm, 56 cm and 66 cm (20 in, 22 in and 26 in).

Internally, the tube is very different from a monochrome tube. At the back end of the neck there are three electron guns instead of the single one of a monochrome tube, and the fluorescent screen, instead of being formed of a smooth paste of phosphor, is composed of a mosaic of phosphor dots. In addition, there is a perforated metal screen carrying a regular pattern of slots located about 1 cm behind the phosphor-dot screen. This is called a 'shadowmask', and it is after this mask or screen that the tube is named the *shadowmask tube* (though it is sometimes called the 'three gun' or 'tricolour' tube).

The phosphor dots are not deposited in a random manner on the screen. They are applied in an accurate and closely controlled process to form small adjacent groups of three vertically elongated dots abreast, one each of red-, blue-, and green-emitting phosphor material, and these groups are in a regular pattern all over the screen surface. For each group there is one slot in the shadowmask. The shadowmask of a typical tube contains about 300 000 slots, which means that the screen consists of almost one million phosphor dots in all, each a fraction of a millimetre across, and almost touching each other. Thus the phosphor makeup of a group is such that the light emitted when the group is bombarded by the beam electrons is of three different colours: one dot in each group emits red light, another green light, and the third blue light. In spite of this, the phosphors of all the dots look the same when the tube is inactive, and it is impossible to tell one colour phosphor from another simply from apperance. The dots are so small anyway that it needs a magnifying glass to see them properly on the inner surface of the shadowmask-tube screen.

The shadowmask is orientated relative to the phosphor-dot screen and the position of the three electron guns in the tube neck so that the three electron beams, one from each gun, pass through a slot in the shadowmask and spread out a little between the shadowmask and the screen and so that the three beams diverge correctly, each on to its intended colour of phosphor dot. Thus the beam from the 'red' gun strikes only the red-emissive dot of a group, the beam from the

136

'blue' gun strikes only the blue-emissive dot, and the 'green' beam activates only the green-emissive dot of the same group. In practice, the beams are rather wider than the slots in the shadowmask, so that a single working electron beam will cover several mask slots and activate several phosphor groups simultaneously.

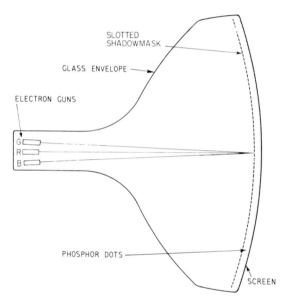

SLOTTED
SHADOWMASK

GLASS ENVELOPE

ELECTRON GUNS

G
R
B

PHOSPHOR DOTS

SCREEN

Figure 6.5. The main features of a shadowmask tube – the phosphor-dot screen, the shadowmask, and the three electron guns

Figure 6.5 shows the main features of a shadowmask tube: the phosphor-dot screen, the shadowmask, and the electron guns at the 'back end'. *Figure 6.6* shows the appearance of a 26 in shadowmask tube of the 30AX type by Mullard Ltd.

Figure 6.7 shows (from above) how the three beams from the guns pass through one slot of the shadowmask to strike separately the red, green, and blue phosphor dots of a group. For the purpose of description, it has to be assumed that the three beams are not being deflected all the time. We shall see shortly that the design of the tube is such that, even when the beams are being deflected together, vertically and horizontally, the shadowmask ensures that they strike only their 'own' phosphor dots.

Figure 6.6. An example of a wide-angle shadow-mask tube with deflection yoke – the 30AX type (Mullard Ltd)

The beam impinging upon the red dots is referred to as the 'red beam' (it is really an invisible stream of electrons) and so on. The guns emitting these beams we will call the 'red', 'green', and 'blue' guns, after the electrical signals fed to them.

Figure 6.8 shows another impression of the working of the shadowmask tube. In fact, the diameter of the beams at the shadowmask is, as we have seen, larger than the slot size, and as a result some of the energy in the beams is dissipated in the mask itself. This warms up the shadowmask, and necessitates high beam current and a high accelerating potential to adequately energise the phosphor dots. EHT voltage is around 25 kV, and since there are three beams to accelerate, plus the energy losses due to the shadowmask, the total beam current on bright pictures can exceed 1 mA. The EHT voltage source must be able to deliver this current without any diminution in voltage.

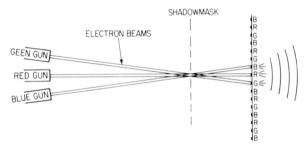

Figure 6.7. The principle of a shadowmask. The beams from the three electron guns pass through one hole of the mask to strike separately their own phosphor dots

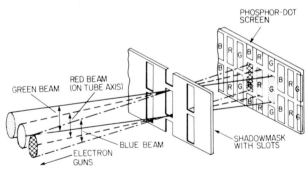

Figure 6.8. Detail of the slot-mask and phosphor screen. Much of the total beam energy is dissipated in the mask itself

Figure 6.8 shows each beam covering only one dot. In reality, each covers several dots of its appropriate colour phosphor. Thus several groups of dots form one scanning spot, the red, blue, and green beams lighting their own colour phosphors in each group.

The electron guns are similar to those used in monochrome tubes, having the same form of heater, electrode system, and electron lens. In some colour-tube types, several of the electrodes in the electron gun are common to al three beams, so only the cathodes are brought out to separate pins on the tube base. The heaters are also internally connected in parallel.

In the modern *in-line* tubes we have been describing, the three guns are mounted in line and abreast at the rear end of the neck, and the central gun (typically red) fires its beam down the axis of the tube to bombard the centre dot of each phosphor group. The blue and green beams, being mounted off centre, would normally suffer serious geometric distortion, and as a result their rasters would not coincide with the red one. The scanning yoke is very carefully designed, however, and its magnetic field is shaped in such a way that errors due to the off-axis sources of blue and green beams are compensated for in the scanning process, and the red, green, and blue rasters overlay each other.

Things were quite otherwise with earlier types of colour tube. For many years, the guns were arranged in delta formation, with a corresponding delta configuration for the phosphor dots, which were circular rather than slot-shaped. Because none of the guns was on the tube axis, and the deflection field was the normal type used in monochrome tubes, a great deal of raster correction had to be applied to achieve convergence, and many wound components and magnets were fitted to the CRT neck, working in conjunction with adjustable correction wavesforms derived from the timebases of the set.

Purity
Let us examine the way in which each electron beam, in spite of being deflected all over the screen surface, excites only its

140

own phosphors. So far as the shadowmask is concerned, the origin of the electron beams is not the gun system but the deflection centre, a point in the middle of the tube neck coincident with the centre of the scan-yoke assembly. The three beams appear to emerge in the same formation (in line abreast) from this point regardless of the direction in which they are heading under the influence of the deflecting field. Let us imagine that we are a 'blue' electron, and that we have been fired from the 'blue' gun and duly arrived at the deflection centre. If we pause here (quite impossible for the electron, but easy enough in our imagination!) and look towards the shadowmask, all we will see, no matter which of the slots we look through, are 'blue' phosphor dots. We cannot see any 'red' or 'green' ones, because the shadowmask is in the way. Beside us is a 'red' electron, and from its point of view every slot in the mask looks out on to a 'red' phosphor dot. It cannot see any 'green' ones – these would be visible to it only if it were able to shoulder aside the 'green' electron beyond it. This means that the deflection-centre-to-mask spacing must be spot-on; this is set in manufacture, either by bonding the deflection yoke to the tube, or by providing precision locating bosses. It also means that anything which alters the apparent source position of the beams will upset beam landing (called *purity*). A stray magnetic field is quite capable of doing this. Magnetism in the shadowmask will also upset purity, and such fields do not have to be very strong – even the normal magnetic field of the earth can upset beam landing.

To counter this, we arrange to demagnetise the shadowmask each time the receiver is switched on. This is easily done by applying a strong alternating magnetic field (sourced from the 50 Hz mains supply) to a large coil wound round the picture-tube bowl. The 'degaussing' field is arranged to decay quickly to zero, effectively removing any residual magnetism from the shadowmask, rim band, etc. Impurity shows up as a 'stain' of another colour when the tube is displaying the output from one gun only: on a blue raster, for instance, impurity may show as a purple stain, caused by electrons in the blue beam spilling over on to the red

141

phosphor dots. In early delta-gun tubes, this could be corrected by adjusting a beam-shift assembly ('shuffle plates') on the tube neck and resetting the deflection centre by sliding the scanning yoke to and fro along the tube neck.

Additive mixing
When a colour tube is properly purified and in correct convergence, a white raster will result when the intensities of the three beams are in the proportions determined by the nature of the screen phosphors and the characteristics of the tube. Although the raster looks white, close examination through a magnifying glass would reveal its composition as minute dots of the three primaries, red, green, and blue, with the green dots appearing to give most illumination, and the blue ones the least, in the proportions given earlier in this chapter (or, at least, close to them) – that is, red 30 per cent, green 59 per cent and blue 11 per cent.

We see white light because our eyes integrate the minute areas of colour into a composite colour (or white when the three colours are correctly proportioned). Indeed, the eyes are unable to discern very small areas of colour in isolation, and they cannot perceive detail in colour. The Y signal thus provides the detail of the picture, using the full bandwidth of a monochrome picture, while the colouring is added by the colour signals. And because detail is not given by the colours, these can be transmitted within a bandwidth considerably smaller than that required for high-definition monochrome television.

Hues and monochrome
Let us see how the shadowmask tube can emit light of almost any colour. We have already seen that by adding the three primary colours in given proportions we can obtain white light. If the three guns of a shadowmask tube are set up in such a way that the beam intensities can be regulated by three separate controls, we can get an idea of the colour capabilities of the shadowmask tube through the selection of primary colours.

142

Let us suppose that we have adjusted the three beams to give a white output; we find that we can increase or decrease this white output by turning the three controls *in step* up or down. Thus, we can go from *peak white to black*, as with an ordinary monochrome tube.

We can get *red, green* or *blue* by turning on the corrsponding gun separately with the other two off. We can get a colour called *cyan* by turning on the blue and green guns without the red one. We can get *yellow* by turning on red and green guns without the blue one. And we can get a colour called *magenta* by turning on the red and blue guns without the green one.

We have thus obtained all shades of monochrome from black to peak white, red, green, and blue, which are the three primaries, and cyan, yellow, and magenta, which are called 'complementary' colours. A complementary colour is produced when a primary colour is removed from white. By removing blue from white, for instance, we get yellow (that is, red plus green *without blue*), which means that yellow is the complement of blue.

We can get all intermediate colours simply by adjusting the relative intensities of the three beams. Orange is produced by a mixture of green and red, with red predominating.

However, there is one more important aspect of the colour display, and that is *saturation*. The complementary and primary hues are called 'saturated' hues or colours. Very few colours in nature are fully satured, so the colour tube needs to display mostly desaturated colours. Desaturation simply implies that some white is added to the saturated hue, and since white is formed of correct proportions of all the primary colours, desaturation occurs when all the colours are being emitted, but with the actual dominant colour well in excess of the others, the excess depending upon the degree of saturation.

It is now possible to see how the shadowmask tube can be used for a colour display in place of the three separate tubes in *Figure 6.3*. *Figure 6.9* shows the setup. Here it is supposed that the tube is provided with timebases and so on, and that each gun is biased to provide beam intensities to give a

143

monochrome picture when the camera is televising a white object, such as a blank white card.

If this white card is changed for one in colour, the display on the shadowmask screen will be in the corresponding colour. It will be seen that the Y output from the camera is not used in this simple example, but this output could be fed to the single gun of a monochrome tube to give the display in

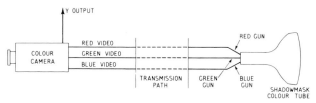

Figure 6.9. Elementary colour-television system using a three-tube camera and shadowmask picture tube. Note the use of three separate transmission channels

grey, even when the camera has a colour input. This illustrates one basic aspect of compatibility. The Y output is used in a proper transmission system, as we shall see.

Transmission

Clearly , the system depicted in *Figure 6.9* is not compatible in itself. It would require three separate transmitters (or line circuits) to provide colour reception. We are left with the problem of replacing the transmission path shown in *Figure 6.9* with a single vision transmitter arranged in such a way that the signals it generates will work a colour set in full colour and a monochrome set in black and white. Moreover, the circuits of the colour set must be such as to produce a black-and-white picture from a monochrome transmission.

Introducing colour information on to a television signal is called *encoding* and extracting it at the receiver is called *decoding*. Over the years, many schemes for encoding and decoding have been evolved. Most, including the German

144

PAL system in use in this country, are based on the American NTSC (National Television Systems Committee) system.

The NTSC system has been in use in America for many years. Different names are used to describe details added or altered by other inventors: PAL stands for 'phase alternate line', for reasons that will later become obvious. PAL is based very firmly on the NTSC system, so it is the latter that must be looked at first.

Encoding NTSC

Figure 6.10 gives a block diagram of the encoding end (at the transmitter) of NTSC. It must be understood that this diagram is in very basic form and that in a practical system there is much more detail. Here the red, green, and blue video signals from the camera are fed to a matrix, whence the

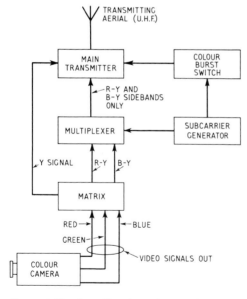

Figure 6.10. Encoding for colour at the transmitter, based on the American NTSC system

correct Y signal (see *Figure 6.4*) is obtained along with the red and blue *colour-difference* signals. These colour-difference signals are derived simply by subtracting the Y signal from the red video signal and, separately, the Y signal from the blue video signal.

The green video signal is not processed in this manner because the green colour-difference signal is recovered later at the receiver simply by the addition of proper proportions of blue and red colour-difference signals (the amount of green light at any time equals the luminance present that cannot be accounted for in terms of red and blue video signals).

So only two colour-difference signals need be transmitted along with the Y signal to derive the three colour signals at the receiver. The Y signal from the matrix is fed straight into the the main transmitter, and since this is the ordinary monochrome signal, the transmitter is now modulated for monochrome, thereby allowing any monochrome set picking up the transmission to produce the colour picture in black and white – a requirement of compatibility.

The red and blue colour-difference signals go to a multiplexer, which is really a special kind of submodulator. This works in conjunction with a subcarrier generator operating at a critical frequency of 4.43361875 MHz, this frequency being chosen to allow the chrominance information to be packed into low-energy bands between the monochrome information of a 625-standard channel and also to avoid certain interference effects.

If we could examine the spectrum of an average TV picture, we should see something similar to *Figure 6.11 (a)*. This shows that most of the luminance information is distributed into 'packets' centred on multiples of line frequency, and that the spaces between these packets are relatively quiet. The diagram also shows that the average TV picture does not contain a great deal of fine detail overall, so that in the high-frequency area, say above 2 MHz, there is a tailing-off of luminance energy.

As we shall see shortly, the chrominance signal consists of a pair of 1.3 MHz sidebands with *no carrier*. If we 'drop' the

146

chrominance signal into the luminance spectrum so that the absent carrier responds to a frequency of 4.43361875 MHz, not only will it occupy the 'quietest' part of the luminance area, but the sidebands, which are in similar line-frequency related packets to the luminance signal, will interleave with the components of the luminance signal and cause a minimum of mutual interference. This 'interleaved combs' effect is shown in *Figure 6.11 (b)*. Hence the choice of the odd subcarrier frequency, which must be controlled very closely indeed at the studio if crawling-dot patterns on the picture are to be avoided.

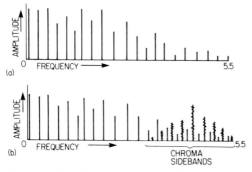

Figure 6.11. 'Packet' energy distribution. Spectrum of a typical video signal, showing the information to be packed into 'bundles' at multiples of line frquency. (a) The luminance signal. (b) The interleaving of the chrominance sidebands into the upper (quietest) part of the spectrum

To modulate two signals (for example, the red and blue signals) on to a single carrier, a technique called 'quadrature modulation' is adopted, and this causes both the *amplitude* and the *phase* of the subcarrier to vary in accordance with the chrominance information. We will return to this later.

After modulation of the subcarrier, the subcarrier itself is suppressed to avoid interference with the luminance signal and the multiplexer delivers *only the sidebands* of the red and blue signals. However, to obtain demodulation of these

147

at the receiver, the subcarrier must be locally recreated in the receiver, at the exact frequency and phase of the original subcarrier at the transmitter. A crystal oscillator at the receiver is used for this purpose and it is 'locked' on frequency and phase by colour-sync signals, called *colour-burst* signals, which consists of short 'bursts' of about ten cycles of signal at subcarrier frequency added to the transmitted waveform during the back porch of the line-sync pulses, as shown in *Figure 6.12*. At the receiver, these bursts are used to establish the correct frequency and phase of the local subcarrier generator, making it possible to demodulate the chrominance signals correctly.

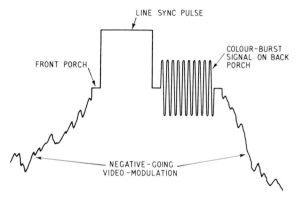

Figure 6.12. The colour burst. This is the reference signal for the decoder, and consists of 10 cycles of subcarrier frequency inserted during the back porch of the line-blanking period

Figure 6.10 shows the subcarrier generator at the transmitter and the colour-burst switch. It also shows that the main transmitter is modulated with the Y signal, the sidebands only of the two colour-difference signals and the colour-burst signals. It is also modulated, of course, by the line- and field sync pulses and the black-level references in the same manner as monochrome transmitter.

148

We have already seen that all television is an optical illusion, and that the eye is fooled into seeing a complete picture when in fact all that is present is a single rapidly-moving spot. This persistence of vision is one peculiarity of the human eye. The colour TV system perpetrates another confidence trick on the viewer: it exploits the fact that the eye cannot perceive fine detail in colour. It would be wasteful to transmit a full-bandwidth signal for each colour, even if the band space to do it were available. This is why we can get away with transmitting a full-definition monochrome picture (luminance signal) and roughly 'painting-in' the colour information on it, in low-definition form. Chroma signals, then, are limited to a bandwidth of about 1.3 MHz, and this is the key to inserting them in a normal 625-line monochrome channel.

The video signal is now becoming quite complex. In a single waveform we have the basic monochrome picture, the chrominance components for a full-colour picture, the burst 'key' for decoding a black reference level, line and field synchronising pulses . . . and we've yet to mention text, data, and test signals. Even the sound information can be carried in the sync pulse, but that's another story.

NTSC decoding

A very simplified NTSC decoding system is shown by the block diagram in *Figure 6.13*. Here the multiplexed signals are detected by the ordinary vision detector, and the Y signal is filtered to a Y or luminance amplifier, which is the same as the video amplifier in a monochrome set. The output of this amplifier feeds the cathodes of the three guns of the shadow-mask tube together.

The multiplexed chrominance signals are filtered into the decoder, in which are the local subcarrier generator, colour-demodulators, colourburst sync, and many other refinements. This gives the red and blue colour-difference signal outputs, which are fed to a matrix to obtain the missing green colour-difference signal. This is obtained by the addition of suitable proportions of red and blue colour-difference signals.

The matrix thus delivers three colour-difference signals into the red, green, and blue colour-difference amplifiers, the outputs of which are also channelled to the cathodes of the shadowmask tube.

The tube guns are adjusted to give the correct intensities of red, green, and blue light from the phosphor dots so that a monochrome display is obtained when a colourless input is applied to the camera at the studio. When the system is balanced in this way, therefore, a colour input at the camera will give rise to colour-difference signals which modify the video signals fed to the picture-tube cathodes, thereby changing the relative intensities of the phosphor dots and resulting in a reproduction of the colour as 'seen' by the camera.

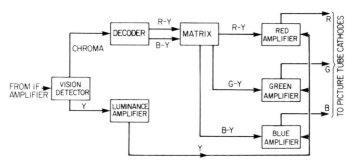

Figure 6.13. Block diagram of a simple NTSC decoding system

If the colour make-up remains the same and only the brightness changes, the Y signal will change and the proportions of colour-difference signals will remain the same. Detail is transmitted by the ability of the Y signal to change very rapidly, corresponding to a bandwidth of about 5 MHz, while the colour is rendered in far less detail by the relatively slower changes in the chrominance signals.

The colour-difference signals, recreated by the decoder in the colour receiver, are added to the luminance signal before application to the picture-tube cathodes. Thus Y, when added to G-Y, gives: (G-Y) + Y = G. The same goes for the R-Y and B-Y signals, so the signals emerge from the final video amplifiers in RGB form. During the transmission of a monochrome picture, no colour-difference signals are sent or received, so the R, G, and B signals consist wholly of Y information. The net result is the same as that shown in *Figure 6.9*, but with the three-circuit transmission path reduced to just one circuit.

Colour sets feature many more refinements than can possibly by described in this chapter, and a book dealing exclusively with colour television – such as *Beginner's Guide to Colour Television* – should be consulted for a more comprehensive description of the working of the colour television system. As with our bandwidth-limited colour-difference signals, only a basic outline can be given here.

The PAL system
The PAL system, which is a German invention, is based on the NTSC system, and all that has been discussed so far is applicable equally to PAL and to the basic NTSC system. However, PAL has a particular refinement concerning the colour stability of the system under conditions of chrominance phase error. To understand this, we must examine the nature of the chrominance signal.

We have learnt that the colour signal consists of the sidebands of a 4.43 MHz signal, which is itself deleted in the encoding process. The amplitude of the chrominance signal determines the strength of the colour-difference signal in just the same way as amplitude modulation of the luminance signal determines the intensity of the monochrome scanning spot. We also have to know the colour (hue) of the signal being transmitted, and this achieved by varying the phase of the chrominance signal. Consider the vector diagram, *Figure 6.14*. The V axis correspnds to R-Y information, and the U axis to B-Y information. These axes are at right-angles to each other, indicating a phase shift of 90°, or one quarter of one subcarrier cycle. The chroma signal itself is represented by the dotted line, which can move to any point in the circle, depending on the hue present. Thus a bright red raster would correspond to a long dotted line pointing due 'north', while a pale-blue raster would find the dotted line extending a little way 'east' along the +U axis. For a green field, the dotted line would need to move into the −U and −V area of the circle, so as to turn down the blue and red guns and turn up the green one. In these circumstances, the dotted line would point 'south-west' in *Figure 6.14*, and the longer it was, the greener the field.

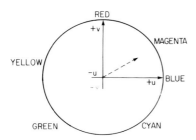

Figure 6.14. Vector diagram for the chrominance signal. The phase of the chroma signal determines picture hue, and saturation is proportional to its amplitude

'Phase' is a relative term. Unless the phase of our dotted line is relative to a steady reference, the whole process is meaningless. In fact, the phase reference is to the subcarrier itself, and as this was deleted in transmission, we have to recreate it in the receiver before we can make any sense of the phase variations of the chroma signal. It is controlled by the burst signal: more of this later.

Having established that picture hue is totally dependent on the phase of the chrominance signal, it becomes obvious that any phase change in the path from the encoder to the decoder (and it's a long path) will upset the hue of the reproduced picture. The fact that the reference signal (burst) is transmitted along with the picture goes a long way to alleviate the problem, because if phase distortion affects the chroma signal, it is likely to affect the burst signal likewise, and if both undergo the same phase error, their relationship is unaffected, and all is well with the reproduced hue. But the chrominance signal, be it remembered, is riding atop the luminance signal, so burst and chroma travel through the system at different amplitude levels, and this is the snag. Under certain circumstances, *differential phase distortion* can occur, in which phase distortion is level-dependent. This upsets the chroma/burst phase relationship, and bad hue changes can result. In extreme cases of multipath reception and non-linear system response, the hue can be badly enough upset, on the NTSC system, to render the sea red and the grass blue. To overcome this, NTSC receivers are fitted with a hue control; it is normally set by the viewer for acceptable flesh tones.

With the benefit of hindsight, the PAL system was designed to overcome this problem. On alternate lines, the red colour-difference signal at the transmitter is reversed in phase (that is, switched through 180°). This is done by a switching pulse operating an invertor at line frequency in the R-Y channel, as shown in *Figure 6.15*. The switching is done line-sequentially, so for one line of signal we get +(R-Y) and +(B-Y), while on the next line the signals transmitted are −(R-Y) and +(B-Y), hence the name 'phase alternation, line'. In the decoder, we feed the chrominance signal to a delay line which effectively stores it for one television line (64 μs). Assuming that one line of chrominance information is very much like its predecessor and successor (and in practice they are), we now have available in the decoder two lines of chroma information simultaneously – the direct and delayed versions. Because one of these was transmitted upside-down, as it were, the errors in it will be equal and opposite to those in the other, and thus will cancel out when they are brought together. Any phase error now shows as a slight saturation change with the hue unaffected, and this is far less objectionable. In practice, the phase error must be quite large to give any appreciable change in saturation.

A PAL set, then, has no need of a hue control, and the only colour control needed by the user is the saturation control,

153

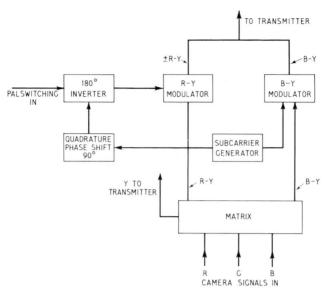

Figure 6.15. The basic elements of a PAL encoder, offering more immunity to phase distortion than NTSC

which regulates the amount of colour-saturation in the picture. Some manufacturers do provide a tint control, but in most sets this has nothing to do with the decoder because it alters the basic white (i.e., the colour temperature of the picture) and is effective on a monochrome display.

PAL identification

Having reversed the phase of the red difference signal, we need to identify which set of lines is phase-reversed. The PAL lines are 'tagged' by advancing and retarding the phase of the reference burst signal in synchronism with the phase reversal of the R-Y signal. Hence the term 'swinging burst'. The fact that the burst phase is advanced by 45° on one line and retarded by 45° on the next has no effect on the carrier regeneration process in the decoder, because the *average* phase of the burst signal is correct.

154

Colour killer and ACC

When tuned to a monochrome transmission, any noise or interference signals present in the decoder will give rise to spurious colour disturbances on the picture, usually in the form of the 'confetti' effect. To prevent this, the decoder is shut down by a colour-killer circuit which operates in the absence of a received colourburst signal.

To avoid saturation changes due to propagation errors and mistuning, automatic colour control is provided. This works in the same way as the AGC system we found in the IF section. As a reference, it uses the burst signal, the only part of the chroma signal transmitted at constant amplitude. The gain of the chrominance amplifier is thereby adjusted to maintain a constant burst amplitude in the decoder, and hence a constant saturation level in the colour picture.

PAL decoder

To complete our account of PAL decoding, a block diagram of a PAL decoder is shown in *Figure 6.16*. The composite chroma signal (in this case, 'composite' means complete with burst) centred on 4.43 MHz is separated from the luminance signal by a suitable filter and amplified in a bandpass amplifier. The signal is further amplified in a delay-line driver stage whose gain is variable by the colour control. The chrominance signal now passes into the one-line delay, after which the PAL-addition process takes place to cancel any hue errors. The R-Y signal undergoes line-by-line inversion by the PAL switch at this point, to restore the 'upside down' lines to normal. here the R-Y and B-Y signals are still in subcarrier-sideband form, so they must be detected before they can be used. This takes place in synchronous detectors rather like those we met in Chapter 4. We have now recovered the R-Y and B-Y signals in the form of video (baseband) signals, ready for the matrixing process; first with each other, to derive the G-Y signal, then all three with the Y signal to produce RGB drives for the picture tube.

So far we have discussed the top half of the diagram in *Figure 6.16*, and this is the section which handles the chrominance signals themselves. The bottom section contains the

155

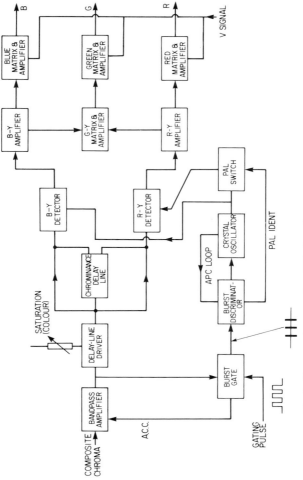

Figure 6.16. Block diagram of a complete PAL decoder

156

reference chain: this is concerned with processing the reference signals for the identification and demodulation processes. After the bandpass amplifier, the burst signal is extracted by the burst gate. As its name implies, it is an electrical gate which is driven by a suitably-timed pulse, and it opens for just the duration of the colourburst. Its output consists, then, of a series of bursts with no chroma information. These are passed on into the *burst phase detector*, which works in a similar way to the line phase discriminator we studied in the flywheel line sync circuit. The oscillator this time is a crystal type, and its frequency and phase are thus held steadily in step with the average burst phase, so that its output represents the regenerated colour subcarrier. It drives the synchronous detectors which recover the R-Y and B-Y signals.

The output from the burst phase detector contains a ripple voltage due to the swinging characteristic of the burst signal. This is extracted and amplified to drive the PAL switch already discussed. We do not want the ripple to upset the crystal

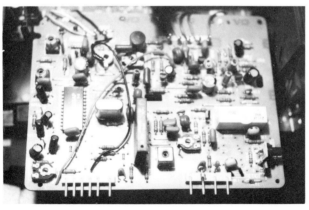

Figure 6.17. A one-chip PAL decoder. The four largest items on the PC panel are, from left to right: decoder IC, PAL crystal, chroma delay line, and luminance delay line. In addition to its colour-decoding function, this panel also handles the luminance-signal processing (GEC-Hitachi Ltd)

oscillator operation, however, and the flywheel effect of the APC (*automatic phase control*) loop ensures that the ripple effect is lost so far as the subcarrier regenerator is concerned.

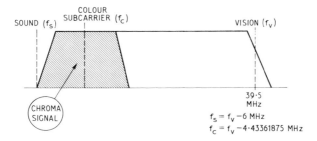

Figure 6.18. The colour signal as present in the IF channel

Many colour receivers use a single chip for the whole of the decoder functions, with a minimum of external components and adjustments. *Figure 6.17* shows such a decoder, which is very much smaller and simpler than early discrete designs and capable of better performance.

The diagram in *Figure 6.18* is the spectrum of a colour-encoded signal as it enters the vision detector.

158

7

Test cards and receiver controls

Since the beginning of the television service in the 1930s, test cards have been broadcast to check the performance of the system. As improvements and system changes have taken place, from 405-line to 625-line and then to colour broadcasting, the design of test patterns has evolved to meet the needs of broadcasters, the TV servicing industry, and viewers. Early test cards were optically generated, first by a camera-and-blackboard technique, then a monoscope (a form of flying-spot scan tube with a test pattern etched into its screen), and finally by a flying-spot scanner (FSS) as described in Chapter 2. Such optically-generated patterns as are still in service (e.g., test card F) are made in 35 mm slide form and televised in an FSS system. Many people in the TV industry feel that the optically-generated card is best from the service engineer's and viewer's point of view because it contains, as in test card F, a 'normal' colour picture, including flesh-tones, which is subjectively more satisfactory than colour bars for appraising the colour reproduction .

As transmission systems become more automated, it gets increasingly difficult, from the broadcaster's point of view, to generate and distribute optically-generated cards. The FSS equipment requires constant maintenance, and the test pattern must be distributed from a central point to the transmitters. This engages expensive transmission lines and equipment at times when it might otherwise be free or used for the exchange of programme material between centres. For these reasons, and to achieve consistency and reliability,

Figure 7.1. Broadcast test cards. Top, Test-pattern F. Bottom, Test-pattern G (BBC)

160

the broadcasting authorities and test-equipment manufacturers, notably Philips, have developed electronically-generated test cards. These have all the important features of the earlier cards, with the advantages of needing no maintenance and of being available on site at transmitting stations. Current optical and electronic patterns are illustrated in *Figure 7.1*: top, optical test card F, and bottom, electronic test card G.

Test cards contain features designed for subjective appraisal of most aspects of signal propagation and receiver performance. In this chapter, shall describe the features of the test card and the way they enable us to adjust the TV set and judge its performance. It is important to emphasise the word 'subjective'. Many aspects of performance can be tested only with special signals and test gear, while in some other respects the test card is a standard too strigent for the system. So we need a separate, special pattern to adjust *convergence* on a colour receiver, while few sets can fully resolve the finest detail on the test card. Receiver design, like most things in life, is a compromise. We cannot expect perfection in the reproduced test card, even on the most expensive receiver; this is not its purpose.

Now that we have seen how the television signal is produced, transmitted, received, and displayed, we should not have much difficulty in understanding how the test pattern checks the receiver, and what effect receiver-control adjustments have on it.

Aspect ratio and scanning amplitude

The aspect ratio (width: height) is 4:3 in the broadcast picture, and except very old types, picture tubes are made to this specification. The borders of the test cards consist of alternate light and dark blocks, called 'castellations', and the width and height of the picture are set so that half of these castellations are visible at the picture borders. When this is done, the central circle should be truly circular, and the white background squares truly square. The height control is usually an internal pre-set, but the form of width adjustment varies with receiver type. As we have seen, picture width is proportional to HT line voltage, so proper adjustment of the power supply should ensure that the width is correct. Where separate width adjustment is provided, it will take the form of a 'tap' system, in which two or three width settings can be selected by a plug-and-socket or soldered wire, or, in some colour sets, a miniature pre-set similar to the height control. What we are doing, of course, when we adjust width or height is altering the amplitude of the sawtooth currents flowing in the scan coils – the stronger the magnetic fields, the further the spot is deflected.

Linearity

If the geometry of the test card is to be good, the scanning spot must traverse the screen at a constant speed. Any acceleration or deceleration of the spot on its journey will result in non-linearity of the pattern, showing as stretching or cramping. Both can be present at the same time, resulting, for instance, in an egg-shaped circle even though the scan amplitudes are correct. An example of this is shown in *Figure 7.2*, in which the field scan is cramped at the bottom and

Figure 7.2. Poor field linearity – bottom cramp, top stretch

stretched at the top of the picture. Further examples of field linearity errors appear in *Figure 7.3*: left, bottom stretch/top cramp, and right, foldover at the picture top, a common effect of maladjustment or faults in the field timebase. To correct these, field linearity controls are provided in the

162

receiver. These take the form of one or two pre-set pots which modify the shape of the field-scanning waveform to achieve good linearity. Inability to obtain a linear field scan by adjusting these controls generally indicates a circuit fault.

Just as we used inductive components to generate line scanning waveforms, an inductive device is necessary to control the horizontal linearity. This takes the form of a

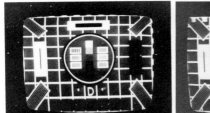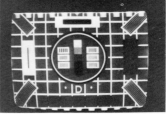

Figure 7.3. Further examples of field linearity problems. Left, top cramp, bottom stretch. Right, top foldover

Figure 7.4. A saturable inductor, used for line linearity control

saturable reactor, pictured in *Figure 7.4*. This is a ferrite-cored inductor with a permanent magnet whose effective strength can be varied by rotating it. The magnet strength sets the point at which the coil's ferrite core becomes magnetically saturated and it 'runs out' of inductance. By inserting this in series with the line-scanning coils, horizontal linearity can be adjusted – the idea is to achieve equal spacing of the vertical white bars in the test pattern. An example of impaired horizontal linearity is shown in *Figure 7.5*.

As there is a degree of interaction between the scan-amplitude controls and those for linearity, it may be necessary to adjust each in turn to get best results – i.e., correct

Figure 7.5. Impaired horizontal linearity – egg-shaped circle and unequal width of side rectangles

aspect ratio with optimum linearity. Often, the picture-shift controls will also need adjusting to achieve the result required.

Picture centring and levelling

Having got the scan amplitudes and linearity correct, we may find that the picture is not centred on the screen, or that it is tilted. In monochrome receivers, picture centring is adjusted by two shuffle-plates mounted behind the scan yoke (*Figure 1.12*). These are adjusted in combination to centre the picture on the screen. The deflection of the scanning spot is carried out, as we know, by the scan yoke, and the electron beam from the gun is pre-deflected by the magnetic fields from the shuffle-plate combination so that it enters the scanning field centrally. What happens if the scanning field itself is tilted? The effect is shown in *Figure 7.6*; it is simply corrected by rotating the entire scanning yoke on the picture-tube neck so that the scanning fields are aligned with the horizontal and vertical axes of the picture-tube screen.

164

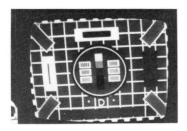

Figure 7.6. Tilted picture

A further distortion of picture geometry may occur, caused by the large deflection angle through which the beam passes and the relative flatness of the screen on which it lands – pincushion distortion, pictured in *Figure 7.7.* In wide-angle monochrome receivers, the deflection yoke is often furnished with magnets around its periphery at the tube-flare end. By manipulation of these, the pincushion effect can be eliminated, but perfection may not be possible.

Figure 7.7. Pincushion distortion. The effect is more noticeable at the sides because the deflection angle is greater in the horizontal direction

All these forms of geometric, yoke-dependent distortions happen in colour picture tubes too. In the cases of the outer electron guns (see Chapter 6), other forms of geometric distortion are also present. We cannot correct such errors as centring and pincushion distortion by simple external magnetic fields, or we would upset the beam landing (purity). Every electron from the 'red' gun has to hit the 'red' phosphors, and they won't if we hang permanent magnets around the tube neck. Geometric distortion, then, has to be countered in more subtle ways in colour receivers. Picture tilt is still corrected by rotating the deflection yoke on the tube neck, but in such tubes as the PIL (precision in-line) type, the tube and yoke are bonded together in manufacture as one unit, so this adjustment is sealed in the factory. In late designs of colour tubes, pincushion distortion is largely compensated for by designing the deflection yoke to have a barrel-shaped magnetic field; in colour sets where pincushion correction is necessary, it is achieved by modulating the line and field deflection currents by suitable *parabolic* waveforms. Such a waveform is shown in *Figure 7.8* for line deflection, and the pincushion-correction controls take the form of pre-set pots and inductors associated with the timebase circuits; these vary the shape and amplitudes of the modulating currents.

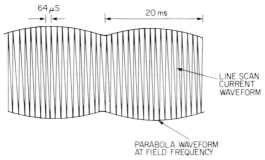

Figure 7.8. Line-scan waveform modulated at field rate for correction of pincushion distortion

Many colour tubes have what appear to be shuffle-plates mounted on the neck, behind the scanning yoke. These are similar to those used in monochrome sets, but they have the function of purity and sometimes *convergence* adjustment. Picture shift is achieved by passing DC in the required direction through the scan coils, and presets (usually heavy wire-wound types) will be found in the timebase circuits of the set to adjust centring, one for each axis.

We have now seen how to check and adjust the geometry of the television picture with the aid of the test cards. But there are many more aspects of the system that they can be used to judge.

Focus

Focus, for instance is revealed not only by how well the horizontal scanning lines are resolved, but also by how sharply the short vertical bars of the frequency gratings are reproduced. The focus, therefore, should be adjusted for the best definition of both. If different focus settings are required for vertical and horizontal detail, a degree of *astigmatism* (caused by the scanning spot being elliptical instead of circular) is indicated, and a compromise is necessary. Most monochrome picture tubes, and some colour types, have a *unipotential* gun, in which the focussing voltage is low and uncritical. In this case, focus adjustment is by a potentiometer or voltage 'taps' and, once set, is unlikely to need further adjustment. The alternative focus arrangement, found on most colour tubes, uses a high (4 kV–7 kV) focus potential, set by critical adjustment of a large potentiometer, often part of the LOPT assembly or closely associated with it.

It is important that the vertical definition bars of the test pattern are properly focussed so that they can be used for their primary purpose, which is to check the horizontal resolution of the system.

Horizontal resolution

In our study of the IF and video amplifier circuits, we saw how the higher frequencies in the video or luminance signal correspond to fine vertical detail in the picture, and how carefully these are preserved throughout the system. The frequency gratings in the test pattern G, reading from left to right, correspond to video frequencies of 1.5 MHz, 2.5 MHz, 3.5 MHz, 4.0 MHz, 4.5 MHz, and 5.25 MHz, and in a well-aligned monochrome receiver all should be visible, though

167

the finer gratings may be reproduced at a lower contrast level due to a falling-off in the response of the video amplifier at high frequencies.

In colour receivers, the definition gratings are not so well resolved. The luminance signal path contains a filter to remove chrominance components (at 4.43 MHz, remember) from the luminance signal, and this has an attenuating effect on any luminance signals within its passband, so that the 4.5 MHz grating will be indistinct. Also, when a colour pattern is being displayed, luminance signals falling within the chrominance passband (about 2 MHz wide, centred on 4.43 MHz, as we saw in Chapter 6) will find their way into the decoder and give rise to spurious chrominance outputs. Thus we see the so-called 'cross-colour' effect on the 4 MHz, 4.5 MHz, and 5.25 MHz gratings, taking the form of blue/yellow and red/green patterning. It is this phenomenon which gives rise to the irritating herringbone cross-colour effect visible on fine detail in ordinary TV pictures, striped shirts and tweed jackets being the among the worst offenders.

Fine tuning

In any receiver, the horizontal definition depends very much on how well the receiver is tuned. Harking back to our description of IF response, it is not difficult to see that when the set is mistuned, the IF output from the tuner will depart from the correct 39.5 MHz. Suppose that slight mistuning brings it to 37 MHz. The IF passband extends only to 35 MHz or so (*Figure 4.6*), so luminance components above 2 MHz are lost off the end, as it were, of the IF passband. Thus no gratings finer than those corresponding to 2 MHz would appear in the display. Fine tuning, then, is carried out by adjusting for finest detail on the test card. If we tune beyond the correct point, the sound and chroma signals become very strong and give rise to coarse dot and herringbone patterns on the test card, so the correct tuning point is for maximum definition consistent with freedom from these effects. At this point, the sound and colour signals should be present and correct.

We have already met AFC (automatic frequency control); where fitted, this tends to mask the correct tuning point by pulling the set into tune when the correct point is approached. If possible, therefore, the AFC should be switched off when tuning is carried out; often the AFC is automatically disabled by a microswitch on the panel or door

Figure 7.9. Tuning controls for self-seeking receiver (Grundig)

giving access to the fine tuning controls. The method of fine tuning varies greatly from model to model – one may encounter a simple rotary control on monochrome portables, press-buttons which rotate to tune on standard sets, and electronic-memory fine tuning press-buttons on sophisticated self-seek receivers. A control panel with the last arrangement is shown in *Figure 7.9.*

Contrast and brightness

Contrast ratio is the ratio between the brightness of the brightest and darkest parts of the reproduced picture. Modern high-brightness picture tubes can achieve a high contrast ratio in low ambient-light conditions. The section of the test pattern concerned with this is called a 'step-wedge', and in test card G it appears as a six-step block below the frequency gratings. The left-hand square corresponds to black level and the right-hand one to peak white, so these are transmitted at 76 per cent and 20 per cent of carrier modulation respectively, and correspond to 30 per cent and 100 per cent of the composite video signal. To adjust the set, the contrast control should be set to, say, two-thirds advanced, then the brightness control adjusted so that the black square is just on the point of extinction. The intermediate steps should then appear as equal increments of brightness. Excessive settings of the brightness and contrast controls may lead to defocusing on peak whites, or a soot-and-whitewash effect where adjacent steps merge into a single black or white rectangle. High brightness and contrast levels are detrimental in other ways too. Ideally, a medium setting of contrast should made, and brightness adjusted to suit, then the ambient lighting reduced as far as possible; apart from giving optimum picture reproduction, this saves two ways on energy.

Streaking, multipath, and ringing

We have looked at contrast ratio and frequency response in the video signal chain with the aid of the test pattern. Another parameter of the set's receiving circuits is the LF (low frequency) response. Where large areas of the televised picture are at the same brightness level, it is important that the IF and video circuits can hold the picture-tube cathode at the corresponding voltage level for the relatively long duration (in terms of scanning time) of the area involved. A lack of LF response, then, is manifest as a change in brightness over an area which is transmitted at constant luminance level. To check this, the test pattern incorporates a black rectangle within a white area – at the top of the circle in *Figure 7.1 (b)*. Poor LF response shows as streaking to the right at each end of the black rectangle, pictured in *Figure 7.10*. Such shortcomings usually originate in the video (luminance) amplifier.

170

The thin black line on a white ground below the rectangle (*Figure 7.1 (b)* again) is called a 'needle pulse' for obvious reasons; in other cards it takes the form of a white line on a black ground. In either case, the purpose is the same: to check the transient and pulse response of the receiver, and for ringing and multipath reception (ghosting). A ghost image may not be very clear on an ordinary picture, but on the test card it will show clearly as a reflection of the needle pulse on its right-hand side and in the contrasting background, making it easy to discern. This test is useful, then, when aligning receiving aerials.

Figure 7.10. Streaking of picture due to poor LF response

Any tendency to *ringing* or *overshoot* in the vision stages of the receiver will give rise to multiple reflections of the needle pulse to its right, alternately positive and negative, and diminishing in amplitude as they retreat from the needle pulse itself. This effect should direct attention to the vision IF alignment and any inductive components in the luminance chain, such as the luminance delay line in a colour receiver. The stripes above the colour bars represent an intermediate frequency between the rectangle and needle pulse and should be reproduced without pre-shoot, overshoot, or 'smudging', providing a further check on video-circuit performance.

Synchronising check

The castellations at the test-card edges serve another purpose besides defining the picture edges. They are transmit-

171

ted as black/white or black/chrominance alternations, depending on the test pattern concerned. The right-hand border serves as a check on the performance of the sync separator. It will be recalled that the sync separator strips off the sync pulses from the composite video signal, and if it fails in this, vestiges of the right-hand edge of the picture may appear in the sync output. This will show on the test card as horizontal displacement of those parts of the picture on the same lines as the lighter-toned rectangles on the right of the test pattern.

Convergence and grey scale

The remaining features of the test pattern are designed to facilitate adjustments and checks of the colour performance of the receiver. Perversely, the first of these is a monochrome pattern! The background of the test cards consists of a white grid on a grey ground, and this is there to provide a check on the *convergence* of the three beams in a colour picture tube. In Chapter 6, we discussed the operation of a colour picture tube, in which a colour picture is built up by superimposing three coloured rasters on the phosphor screen. If these three rasters do not exactly coincide at all points of the screen, the red, green, and blue pictures will be out of register with each other, and colour fringing will result. The central cross, a feature of all colour test cards, is there to assist the setting of the *static convergence*, which is adjusted at the tube neck by means of permanent magnets in picture tubes where this adjustment is provided. The idea is to ensure that the three rasters are coincident at their centres; this is checked by the central cross, which should be truly white without any sign of colour fringing.

Towards the picture edges and corners, another factor comes into play. It is more difficult to achieve convergence in these areas because the three rasters come from three distinct guns in the tube neck, and each undergoes a different type of geometric distortion. Thus colour fringing towards the screen extremities is more common, and the means of correcting it depends on the type of tube. Early designs had a bank of 14 or so *dynamic* convergence controls, with which carefully shaped correcting currents were fed through a *convergence yoke* on the tube neck. More modern colour tubes have correction for convergence built into the scan yoke design, so in the 30AX tube described in the last chapter, no convergence controls are necessary or provided. When setting up tubes in which adjustments for dynamic convergence are provided, the test card is not ideal for the purpose, a locally-generated crosshatch pattern being far more suitable.

A fundamental feature of the colour TV system is that the colouring components are added to the basic monochrome image to form the full colour picture. If the monochrome picture is not correct at the outset – maybe biased towards one colour or another – the hues will be wrong in the full-colour picture, and black-and-white pictures will be tinted. We have to ensure that the light output from the 'red', 'green', and 'blue' phosphors is correctly proportioned so that, with no colouring information present, the colour-tube display is truly grey at all levels of brightness from black to peak white. The idea is to set the three electron-gun cutoff points so that they coincide; then dark greys will be reproduced as such without any tendency to tinting. To ensure a true white in the bright (highlight) areas of the picture, the cathode drives are made adjustable; this forms in effect a contrast control for each individual gun and allows matching of the 'red', 'green', and 'blue' highlights to give a true white, known as 'illuminant D', which we met in Chapter 1.

The test card is ideal for checking and adjusting grey-scale tracking, though it is best first to remove the colour from the display by turning down the saturation, or detuning. Grey-scale tracking is carried out by adjustment of pre-set pots associated with the video amplifiers. First, the cutoff point is set by trimming a pot which sets the DC level of the gun involved; this is carried out on one or more guns to establish neutral colour on the darkest square of the step-wedge. If necessary, the 'drive' pots can then be adjusted to provide a true white on the highlight – RH square of the step wedge.

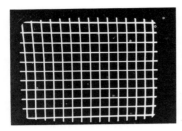

Figure 7.11. Crosshatch pattern for evaluation and adjustment of convergence

When grey-scale tracking and convergence have been set up, very little colour should be visible when displaying a monochrome pattern. *Figure 7.11* shows the special cross-hatch pattern used for checking and adjusting convergence. Another pattern, similar but consisting of a matrix of dots, is sometimes used.

Colour features of the test card

The remaining sections of the test card carry chrominance signals and are concerned with the operation of the decoder section of a colour set. In monochrome receivers, of course, they appear as various shades of grey. Let us start with the colour bars, which appear at the top of test card F and towards the middle of the electronic test patterns. They consist of the three primary colours and their complementaries (i.e., yellow, cyan, green, magenta, red, and blue, reading from left to right). They are described in detail in the companion volume *Beginner's Guide to Colour Television*. For our purpose here, suffice it to say that they enable us to correctly set the saturation (colour) control – the bars are transmitted at 100 per cent saturation, so that the colour setting is right when the 'blue' and 'green' phosphors just become extinguished in the red block of the colour-bar.

On some test cards, the border castellations are heavily saturated in red, blue, and yellow. The edges of the picture are adjacent to the transmitted colour burst, which, as we saw in the last chapter, is used as a subcarrier phase reference in the decoder. The saturated border blocks, then, are inserted to check on the subcarrier locking performance and the correct timing of the burst gating pulse. If the latter is too late, for instance, the subcarrier oscillator will 'see' some of the left-hand border blocks as part of its synchronising signal and try to lock to them instead of the burst. Colours in affected lines will have the wrong hue, betraying the defect as streaks of incorrect colours in line with the coloured blocks on the left-hand border.

The electronic test cards also feature a red rectangle on a yellow ground. This test gives an interesting insight into the operation of a colour receiver and is worth pursuing a little further. We have seen that the luminance signal is handled throughout the set in wideband (0–5.25 MHz) form, whereas the bandwidth of the chroma circuits extends only to 1 MHz or thereabouts. What happens when a sudden change of level in the picture causes the luminance and chrominance

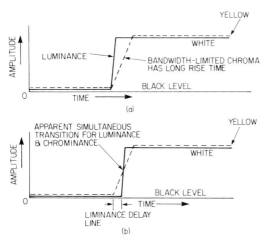

Figure 7.12. How the luminance delay-line compensates for the differing bandwidths of luminance and chrominance channels. (a) The cause of misregistration of chroma and luminance. (b) The delayed luminance signal is back in register with the partnering chroma transition

signals to change level simultaneously? Imagine a transition from black to yellow. This involves a sudden increase in luminance drive and a simultaneous increase in R-Y and G-Y drives. The waveform is shown in *Figure 7.12 (a)*. We can see that the Y signal, because of the wideband characteristic of its amplifier, reaches its new level almost immediately, signifying a fast 'risetime', whereas the chrominance signal, constrained in a narrow bandwidth system, takes longer to rise to full amplitude. This bodes ill for the colour picture. It will have the effect of printing the colour information slightly to the right of the corresponding luminance, because of the time lapse (effective delay) of the chrominance system response. *Figure 7.12 (b)* shows how this is overcome. The luminance signal is passed through a wideband delay line whose delay is chosen so that the luminance signal emerging from it coincides with the partnering chroma signal, and the 'painting-in' of the colour information is in register on the picture-tube screen. To check that the timing is right, the yellow-red-yellow colour-fit rectangle is chosen to show up any errors. The red rectangle should fit snugly between its yellow companions. The coloured vertical rectangles on each side of test card G are intended to check decoder performance at various specified chrominance vector angles, and do not appear on the other test cards.

We have covered all aspects of the test cards now, and no doubt the reader will agree that they are cleverly designed and thoroughly useful in checking every aspect of receiver performance. Broadcast test cards usually carry identification of the transmitter at the bottom.

Auxiliary controls

We have discussed the setting of the main (user) controls in connection with broadcast test cards, and many of the secondary controls inside the set. Let us briefly examine those that we have not touched upon. The hold controls set the free-running frequencies of the line and field timebase oscillators, and the effect of loss of hold for field and line are illustrated in *Figures 7.13* and *7.14* respectively. Simultaneous loss of both suggests a lack of sync pulses from the separator. As the hold controls are adjusted, the sync circuits will pull the timebase oscillators into lock as the correct frequency is approached and thus mask the correct setting. They should be set to the middle of the range over which they have no effect on the picture. Sometimes, a line phase control is encountered. This adjusts the timing of the locked line scan

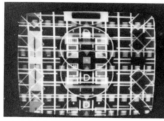 

Figure 7.13. Loss of field syn-chronisation

Figure 7.14. Loss of line syn-chronisation

and has the effect of centring the picture within the raster. It should not be used indiscriminately as a shift control but set by reducing the width and adjusting for an equal margin on each side of the test card, relative to the raster. When this control is incorrectly set, one edge of the picture will tend to fold over as in *Figure 7.15*. Any sideways displacement of the picture when correctly phased and the width restored should be corrected by the horizontal shift control, if fitted.

Most sets have a preset labelled 'set HT' or 'set EHT'. This governs the operating voltage of the set and should only be adjusted in accordance with the manufacturer's instructions, using an accurate test meter, as virtually all other adjustments in the set are dependent on the correct setting of the HT line. We do not propose to go into decoder and alignment procedures, as they are totally dependent on the type of circuit used and are really the province of the service engineer with suitable data and test equipment. One other control should be mentioned, however, and this is the *tuner*

AGC or *delay* control, taking the form of a small resistive preset in the IF or tuner areas. We met it in Chapter 4 and found that it is there to 'bias back' the tuner and prevent RF overloading. Normally, it is left in the full-gain position, but it may be backed off if the incoming RF signal is so strong that

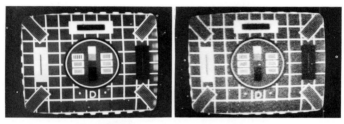

Figure 7.15. Line foldover at the right-hand side, due to a phasing error in the line synchronisation

Figure 7.16. Effect of noise on the picture (snow effect)

cross modulation occurs, showing as patterning on vision and buzz on sound. If this control is retarded on a normal-strength signal, the picture will be snowy and grainy as a result. The effect of this on a received test card is shown in *Figure 7.16.*

8

Closed-circuit television and video recording

Previous chapters in this book showed how a television signal is generated, transmitted, received, and displayed, and how colour is added to the basic signal to build up a full-colour picture. It is now time to turn from broadcast television to the use of television systems on a local basis, often for purposes other than entertainment.

A sound signal is relatively easy to transfer between two widely-separated points, for two reasons. One is that its bandwidth is limited: for speech purposes, we can get away with as little as a 3 kHz bandwidth, and the most common example of relatively low bandwidth speech transmission is the telephone system. Even high-quality sound needs no greater bandwidth than 18 kHz, small when compared to TV systems. The second advantage of audio signals, from a transmission point of view, is that at any instant, a single voltage conveys all the information available: an audio waveform can be at only one level at one time, and there is no need for any dissection or analysis of the signal at each end of the line, as is carried out by the scanning systems of television cameras and picture tubes. Provided a memory bank is incorporated in the receiver, we can in fact send data, in the form of words and figures, and also simple pictures called 'graphics', over a low-bandwidth telephone line, but this calls for very special equipment at the sending and receiving ends, as we shall see in the next chapter.

Closed-circuit television

Closed-circuit television (CCTV) systems differ only from broadcast setups (from an engineering point of view) in that the link from the camera or studio to the receiver or *monitor* is by cable rather than via an RF carrier in space. A monitor is basically the same as a television receiver but shorn of tuner and IF stages, so its input – rather than a modulated RF carrier – is a composite video signal like that illustrated in *Figure 2.7*. The standard distribution level for video is 1 V peak-to-peak at 75 Ω impedance, so in the composite signal the sync pulses occupy the first 300 mV, and black to peak white the remaining 700 mV. The output of TV cameras conforms to this standard, so the simplest form of CCTV system consists of a single camera linked to the monitor by a length of coaxial cable. Because CCTV applications are usually more exacting than domestic TV reception, and because cost is not such a dominating factor in its design, the performance of a monitor is usually far better than that of a commercial TV receiver, and better geometry, video bandwidth, and reliability may be expected from a purpose-designed monitor. They come in screen sizes similar to those of ordinary TV sets, and in both colour and monochrome types. In some applications, the picture performance of a full-blown monitor is not justified on technical or financial grounds: for security surveillance and similar non-critical situations, an adaptation of an ordinary TV set, or even a standard TV set fed from a UHF modulator (a device like a tiny TV transmitter working on an unused channel and feeding the TV aerial socket via a cable), is quite adequate and financially very attractive. A typical colour monitor is shown in *Figure 8.1*. Some designs of domestic TV receiver have a video and audio in/out facility. This combines the advantages of receiver and monitor, and with such a signal source as a domestic VCR or single-tube colour camera, the quality of the input signal is the limiting factor; a professional monitor would offer very little improvement in picture quality over that of a set of this type working at video frequency.

Scanning standards for CCTV systems usually follow the broadcast TV standard for the country in which the equipment is used. Thus in the UK, 625/50 is the norm, and in the USA, 525/60. This is done for several reasons: components for use at these frequencies are very widely available, and the system is compatible with off-air or recorded broadcast material. For some applications, far greater definition than is available from broadcast scanning standards is required, and then special equipment is designed and used. One such case

Figure 8.1. Professional TV monitor. The 37 cm tube has four times as many phosphor dots per unit area as a conventional shadowmask tube, offering a resolution capability of up to 1 300 lines per picture width (Barco Electronic nv)

is medical education, in which perhaps a 1000-line picture is required, reproduced in colour on a number of 70 cm monitors, so that medical students can easily follow every detail of a surgical operation; this calls for such diverse design requirements as a 20 MHz luminance bandwidth and a sterile

181

camera. Interlaced scanning is not essential in CCTV, and a simple system may have independent field and line oscillators, so that the scanning lines of the second field will be traced at random over those of the first. This technique, though it sounds haphazard, is subjectively quite acceptable, although it is now falling into disuse with the advent of colour closed-circuit television (C-CCTV) systems and the availability of cheap IC sync pulse generators, all of which provide fully-interlaced scans.

The camera

Whereas a monitor is usually superior to a broadcast-receiving set, the camera used in the majority of CCTV systems is inferior in performance to the studio cameras used by broadcasters. This is certainly true of domestic and semi-professional cameras, although for *electronic news gathering* (ENG) on location, the broadcast authorities often use a simple portable colour camera very like the types we shall describe.

The heart of any TV camera is its pickup tube. Vidicon types are usually found in CCTV systems. These were described in Chapter 2; as we saw, they need a good deal of peripheral circuitry to operate them, so timebase generators, video amplifiers, and so on are also present in the camera. For domestic cameras, where the output is usually fed straight into a monitor, TV set, or VCR, the sync pulse generator is also built into the camera, so that it will operate with just two connections – power in and composite video out. Such cameras are illustrated in *Figure 8.2* – at the top is a monochrome type, and below are portable colour cameras with zoom lenses.

In a more complex studio system, two or more cameras may be in operation, and their outputs will be required to be blended, superimposed, faded, or 'wiped'. This simply cannot be done if the cameras are not singing from the same hymn-sheet, as it were, so their scans must be synchronised. This is done by feeding them from a central sync-pulse generator which may also insert the syncs on to the studio video output line. This is known as *slaving*; *Figure 8.3* makes

Figure 8.2. Domestic TV cameras. Top, a monochrome type by Ferguson. Bottom, two typical colour cameras by Panasonic, also showing the controls of the partnering portable VCR

183

it clear. It is a pity that many commercial cameras for domestic use do not have slave facilities.

In the chapter on colour television, we saw that broadcast cameras use as many as three or four camera tubes. Such cameras are capable of providing immaculate colour pictures, and for some exacting C-CCTV work these are the only types capable of doing the job. For the sake of portability and

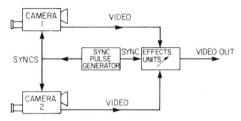

Figure 8.3. 'Slaving'. Synchronisation of the scans of two TV cameras

economy, ENG, semi-professional, and domestic colour cameras utilise two-tube or one-tube ciruits, the latter being preferred as they save on precision optical components and avoid the difficulties of registration (i.e., making the images from the pickup tubes overlay on the reproduced picture). How does a one-tube colour camera work? There are two systems and both involve 'striping' of the faceplate or target of the vidicon. In other respects, the tubes are similar to their monochrome counterparts.

In the stripe-faceplate system, a striped colour filter is bonded to the faceplate of the tube. The stripes are at different angles to the scanning lines, and they thus give rise to different frequencies in the electrical output of the tube (see *Figure 8.4*). Riding on the normal video output of the vidicon, then, are two separate HF signals, whose frequencies correspond to the strip spacings (at a given scanning rate) and which are modulated with R and B information, which may be filtered and demodulated to form colour or colour difference signals. As we saw in Chapter 6, a definite

184

mathematical relationship exists between Y and R, G, B signals, and the G signal may be derived in a matrix from the available R, B, and Y signals.

In domestic cameras, the colour signals are PAL-encoded before being passed out of the camera, as is essential for compatibility with video recorders and TV sets. This makes things simple for the user, if not the camera designer, but the inevitable slight inaccuracies in encoding and decoding, and the (fortunately diminishing) practice of modulating on to UHF and subsequent processing in the tuner and IF stages of

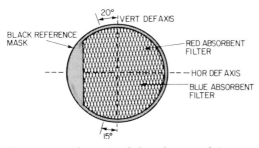

Figure 8.4. The striped faceplate used in one form of single-tube colour-camera system

a receiver, mean that colour reproduction is not so good as it might be if the signals were passed through the cables in simple RGB-Y form. A basic block diagram of a stripe-filter camera appears in *Figure 8.5*. This works from 12 V DC and operates to PAL specifications. It is usually partnered by a portable battery-powered video recorder so that the whole outfit can be taken on location and used like a cine camera, but with the advantage that instant on-site playback is available via the 1 in monochrome monitor/viewfinder built into the camera.

When we consider that scanning, sync generation, video processing, encoding, and all the peripheral circuits associated with these functions are encompassed in the small camera case, we see that a high degree of miniaturisation and

185

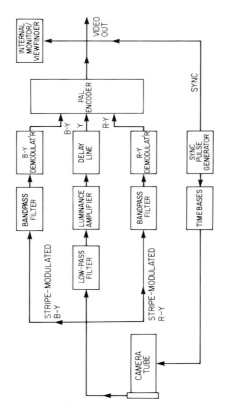

Figure 8.5. Simplified block diagram of single-tube colour TV camera

186

Figure 8.6. Dense packing is necessary to accommodate all the electronics in the small camera case. An internal view of a domestic TV camera (JVC)

high-density packing is necessary inside a colour TV camera. *Figure 8.6* gives some idea of what has been achieved, and we can only congratulate the designers (and repairers!) of this equipment.

CCTV distribution
In video form, CCTV signals on the 625/50 system embrace a bandwidth from virtually DC to 5.5 MHz, as one would expect from our discussions of the broadcast system. In this form, they can be sent down a coaxial cable, and with low-loss 75 Ω cable, a maximum distance of about 500 m may be tolerated without serious loss of picture quality. At greater distances, reception is marred by HF losses and phase distortion in the cable, and it becomes necessary to modulate the picture signal on to a VHF carrier. With suitable equalisation, distances approaching 3 km can be achieved, and still greater distances with amplification and equalisation at suitable intervals along the line.

187

CCTV applications

As time goes on, and the cost of employing personnel increases, more applications are being found for closed-circuit television. Cameras are now found monitoring traffic at road and rail crossings, in cities and on motorways; for these applications, remote control of pan and tilt is often provided so that a large area can be surveyed.

Security is another job suitable for CCTV. A number of cameras can be linked into the system, and their outputs monitored at a central point. For night-time observation of premises, an alarm can be electronically triggered if any picture's video waveform changes, signifying an intruder or disturbance. In stores and other public places, the TV pictures can be constantly recorded and erased, so that by stopping the record/erase process, any crime committed can be captured for use as evidence in a court of law. Prisons, too, can have cameras installed at vantage points around the boundaries.

To accomodate audience overflow at meetings and lectures, microphones and cameras can be installed to convey the proceedings to those without. In hospitals and airports, at railway stations and electricity stations, CCTV has its place. For industrial use, cameras have been developed which are unaffected by gas, water, X-rays, and pressure, so that they can be used in environments which are inaccessible to human observers. Similarly, miniature TV cameras are available which can be passed along tubes or pipes, and with the use of thin fibre-optic light pipes, the medical man can use CCTV to explore lungs and other internal organs. It is impossible to list all the current applications for CCTV in this chapter, but the above examples give some idea of the diversity of applications for this medium.

Closed-circuit television finds very wide application in education. Apart from its use in seminars, conferences, and refresher courses for professional people, it is a useful tool in universities, colleges, and schools. Some years ago, a comprehensive CCTV system was set up by the Inner London Education Authority which embraced a large number of schools and colleges within its area. Signals were distributed at VHF, rather like the relay TV system described in Chapter 3. With the advent of relatively cheap VCR facilities, however, the system has now been abandoned and dismantled. Educational TV production goes on as before, but rather than being piped to schools on the network, programmes are now distributed in the form of video cassettes, and replayed as and when required in the classroom.

Many schools, most technical colleges, and all universities have internal CCTV systems, working at video frequency or VHF, depending on the size of the network. Except for very special purposes, such as the above-mentioned surgical operations, the systems work at 625/50 scanning standards with PAL colour. Apart from distribution of live and recorded broadcast educational programmes, other material, locally generated, can be distributed over these networks. The extent of a network depends on the size and resources of the institution concerned, and cameras, caption generators, flying-spot slide scanners, or even electron microscopes can be used.

Videotape recording

If we can fairly easily record sound on magnetic tape, why not pictures? From the earliest days of television, a means was sought of recording and replaying picture signals. As long ago as 1935, a form of video recording was available, taking the form of a gramophone record marketed by the Baird company. When used in conjunction with the Baird mechanical television receiver, a few minutes of shadowy pictures could be replayed.

Since about 1957, there has been progressive development in video tape recording. In fact, this kind of recording has virtually taken over from film in many television applications. Videotape recording has very much in common with the recording of sound only on a special plastic tape, one side of which is coated with an iron-oxide material, called 'magnetic recording tape'.

189

The electrical sound and picture signals are first translated into magnetic field changes across a very narrow gap in the recording head, and these changes (in intensity and magnetic polarity) are then permanently recorded on the oxide side of the tape in the form of magnetic patterns. It is possible to erase these patterns simply by destroying the magnetism, and this makes it possible for the tape to be used over and over again.

The scheme has the great advantage over film that immediate recording and replay are possible without the need for processing in any way. The principles of magnetic sound recording apply to video recording, and there is virtually the same degree of flexibility. Videotape recording is more exacting than sound recording, mainly because the composite video signal is much more complicated than a sound signal.

It was shown in Chapter 1 that video signals are composed of component frequencies extending to several MHz, and these must be adequately recorded (and replayed) for well defined videotape recording. Sound signals work in a frequency spectrum up to about 20 kHz, but for vision the spectrum has to be extended to 5 MHz and beyond in certain cases where reproduction with very high definition is required.

Recording principles
The changing magnetic field across the gap of the recording head creates on the oxide side of the tape small magnets corresponding in length to the frequency of the signal and the tape velocity and in strength to the amplitude of the signal. Two magnets of opposing poles are produced for each complete signal cycle, and as the frequency increases, so the time taken by a signal cycle reduces. This means that a constant tape velocity results in the recording of magnets of reducing length with increasing frequency. However, as the tape velocity is increased, so also is the length of all the recorded magnets in proportion.

Very short magnets cannot be recorded because they tend to demagnetise themselves as soon as they are made because

190

of the closeness of their north and south poles. Even if short magnets are held by the oxide, they will give a response on replay only when the active gap of the replay head is narrower than their effective length. So the replay-gap width must be less than the length of the recorded magnets in order to 'define' them, as it were. For videotape recording, therefore, where the component frequencies to be recorded are very high, the tape velocity must be increased well above that required for good sound recording. The overall response is also enhanced by the use of replay heads with very small gap dimensions. But there is a physical limit so far as the head gap is concerned, and even the most diminutive of gaps still demands a tape speed well in excess of that for sound recording.

Good audio recording heads respond at a rate of about 800 Hz per 1 cm/sec of tape velocity. One can thus expect a response up to about 15 kHz at a tape speed of 19 cm/sec with equalisation. It is possible to improve on this by the use of heads with smaller gaps, but then the rate of head wear is increased, although this is to some extent countered by the recent developments in ferrite heads. To record a television signal on tape, it is obvious that a high writing speed (i.e., tape-to-head velocity) will be required, and rough calculation suggests that to record a 5 MHz-wide TV signal with practical tape heads, the writing speed would need to be 15 m/sec or 34 mph. This, of course, is quite impossible, both mechanically and because an enormous amount of tape would be needed for a programme of useful length. A way had to be found to increase the tape-to-head speed while keeping the reels turning at a practical speed.

Transverse scanning
The key to achievement of a high writing speed at low tape velocities is to move the head over the slowly-passing tape. The head is made to rotate and scan the tape as a television image is scanned on a picture-tube. Thus a high writing speed is attained, and tracks are laid down across the tape – hence the name 'transverse scanning'. Professional machines used by broadcasters generally have four video heads

191

mounted on a drum whose axis is parallel to the tape, as shown in *Figure 8.7*. Each head lays down about 16 television line waveforms in its sweep from one side of the 5 cm wide tape to the other. So to record a complete television field of 312½ lines, something like 20 head-passes are required (five revolutions of the four-head drum). If we could see the magnetic tracks on the tape, they would look something like those drawn on the tape in *Figure 8.7*.

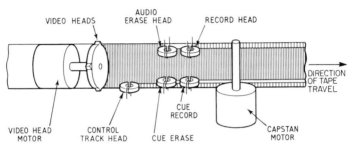

Figure 8.7. The professional transverse-scan recording format. The heads scan the tape across its width to record and replay

The transverse-scan system is capable of excellent performance, but its expense precludes its use in semi-professional and domestic situations. Machines for these applications also use a rotating head drum, but the format here is called 'helical scan', and two video heads are used, mounted on a spinning drum (*head drum*), around which the

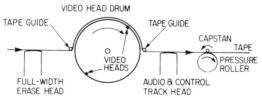

Figure 8.8. The helical-scan recording system using two video heads. This shows the 'U' wrap, so called after the path round the head drum

192

tape is wrapped while being slowly drawn along. The idea is shown in *Figure 8.8*, where it can be seen that the tape is wrapped round rather more than half the head-drum circumference, so that with two heads mounted diametrically opposite on the drum, one is always in contact with the tape. The video head drum is mounted at an angle relative to the tape, and the path of one head across the tape is shown in *Figure 8.9 (a)*, and the resulting track formation in *(b)*.

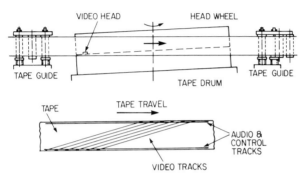

Figure 8.9. (a) The path of the video heads across the tape surface. (b) The resulting slant-track formation. In the Philips 2000 system, the flip-over cassette system scans only half the tape width during a recording, as shown in (a). Other systems use the full tape width during each head scan

The head drum rotates at 25 revolutions per second, so one revolution of the head drum occurs in 40 ms. Each head is in contact with the tape for half a revolution – i.e., 20 ms, corresponding to one field period of the television signal. Thus each slanted track in *Figure 8.9* contains one television field, consisting of 312½ lines, laid end-to-end along the track. The longitudinal tracks at the tape edge contain audio and servo-control pulses – more of these later. So far so good, but there are several reasons why we cannot run the standard video signal straight into the recording heads, and it's now time to see why.

Dynamic range and equalisation

The *dynamic range* of a tape recording system is the ratio of the highest level of signal which can be recorded (limited by *magnetic saturation* of the tape itself) to the lowest usable signal which can be recognised amongst the noise in the system. This ratio is about 60 dB. If we replay a tape-recorded signal, be it audio or video, the output from the replay head is proportional to the *rate of change* of the magnetic flux seen by the head. For a constant-level recording, then, the rate of change will be low for low frequencies (long wavelengths) and high for high frequencies. This means that head output is directly proportional to recorded frequency, giving rise to the curve shown in *Figure 8.10*, where it can be seen that a doubling of frequency (known as one *octave*) results in a 6 dB increase (×2) in replay head output.

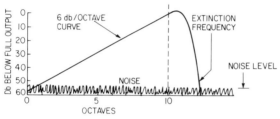

Figure 8.10.   Output voltage-frequency relationship for a magnetic recording system

In a practical tape system, the upper frequency limit is determined by head gap and tape speed, and when the recorded wavelength becomes comparable with the width of the head gap, the response drops away very rapidly towards the *extinction frequency* (*Figure 8.10* again) where one complete magnetic cycle on the tape exactly occupies the head gap, and cancellation of the signal occurs. We can see that even with heavy *equalisation* (to compensate for the 6 dB/octave curve), we are limited to a recording range of about 10 octaves, which is all we can accommodate between the intrusion of noise at the LF end, and the onset of roll-off at

194

the HF end. A range of 10 octaves is sufficient for audio purposes, where it will permit a frequency response of 20 Hz–20 kHz. Video signals, as we have seen, have an important DC component. We can manage without this, but for good colour pictures, we need to be able to record and replay down to 6 Hz and up to 5.5 MHz, representing nearly 20 octaves. Even the less exacting requirement of domestic recording (25 Hz–3 MHz) calls for a 17-octave range, obviously impractical for direct recording.

Somehow, then, we have to reduce the octave range of the video signal before committing it to tape. The problem is overcome by converting the signal to FM. Thus the amplitude variations are converted to constant-amplitude frequency changes. Let us assume that we choose a centre frequency of 3.7 MHz. Black level may take the carrier down to 3.4 MHz,

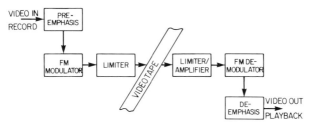

Figure 8.11. Record/replay processing for the FM recording system

with peak white corresponding to, say 4.4 MHz. One side band of the FM signal only is recorded on the tape, and with a typical extinction frequency at 5.5 MHz, this is necessarily the lower sideband. This will extend down to about 1 MHz, so we have now accommodated the luminance signal in a frequency band of 1–5 MHz, corresponding to a recording range of less than three octaves. We have also won another advantage with our FM system. Because of the nature of the tape and variations in tape-to-head contact, it is difficult to maintain constant recording amplitude, and the relative immunity of the FM system to amplitude variations is indispensable for

195

faithfully recreating the video signal. To get our video signal on and off the tape, then, we need, apart from the expected pre-emphasis and equalisation, an FM modulator in the video record path to the heads, and a corresponding demodulator in the replay signal path. *Figure 8.11* shows the idea in block diagram form.

Recording the colour
We have seen that the video response extends to 3 MHz or thereabouts for domestic machines. This leaves the colouring signals, based on 4.43 MHz, somewhat out in the cold! We know from Chapter 6 what liberties are taken with the colour signals in the broadcast system, and for domestic VCR applications, the colour signal undergoes further indignities.

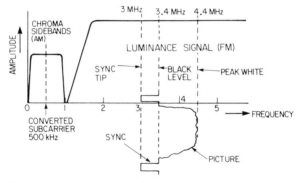

Figure 8.12. Spectrum of the signal on videotape

It is extracted from the composite signal and transposed to a low frequency of about 500 kHz. In this form (still AM and PAL-encoded), it is recorded on the tape along with the wideband FM luminance signal. This acts as a convenient recording bias for the AM chroma signal; this universal technique is known as *colour-under*. During replay, the chroma signal (bandwidth-limited by the recording process, but still subjectively acceptable) is restored to 4.43 MHz by a

196

heterodyning process and recombined with the now demodulated luminance signal. The spectrum of the signal as it appears on the tape is shown in *Figure 8.12.*

Servo control
For this video-recording scheme to work, it is obvious that the head-drum drive and tape-transport mechanisms must be mechanically impeccable and capable of close control. This is necessary to ensure that the video heads accurately lay down tracks on the tape (this confers compatibility with other

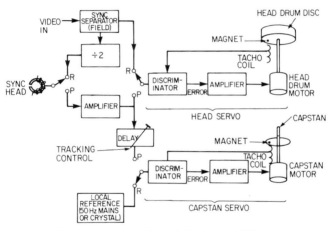

Figure 8.13.   Servo systems for a helical-scan VCR

machines of the same type) and, even more important, that correct *tracking* is achieved on replay. Tracking is the accurate scanning of each track of recorded video by the heads during replay, and a moment's thought will show that this depends on the 'phasing' of the recorded tracks relative to the helical-head path over the tape. This phasing is achieved by *servo*-control of the capstan motor. We met electrical servo loops in the flywheel discriminator circuits of TV receivers (Chapter 5). The servos used in VCRs contain a

197

mechanical system within the loop, so we have real flywheels as well as their electrical analogies.

Let us look at the servo systems used in a typical VCR. *Figure 8.13* shows the two basic loops used to control the capstan (tape transport) and head speeds. Taking the upper section first: on record, the head-drum speed and position are governed by the timing of the incoming field sync pulses. This ensures that each recorded track is exactly one TV field long and that the field sync pulse is at the beginning of each track. At the same time, 25 Hz reference pulses are being laid down on the control track (edge of tape) by a stationary head, independent of the video heads. During replay, these are used to control the video head drum position (phasing) so that each head starts its scan of the tape just before the field sync pulse arrives.

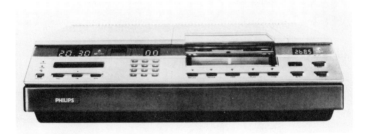

Figure 8.14. VCR using the 2000 format. This machine offers four hours recording time on each side of the cassette (Philips)

We must now look at the bottom of *Figure 8.13*, which represents the capstan servo. This governs the rate at which the tape is drawn past the heads; during record, the capstan motor is controlled by a reference signal derived from an accurate and stable source, either the incoming mains frequency (50 Hz) or a crystal oscillator. This ensures a constant rate of tape speed past the heads. On replay, it is necessary to adjust the position of the tape laterally (with reference to the head drum) so that the video heads scan along the centre of each video track on the tape. To achieve this, the capstan

198

Figure 8.15. Inside a VHS machine. Top, a general view. Bottom, a closeup of the video head drum and tape path (Mitsubishi)

motor is controlled by the 25 Hz reference pulses laid down during the record process, so that a fixed relationship is established between the tape-track and head-drum positions. On many machines, a tracking control is provided, which the viewer can adjust for optimum tracking and minimum picture disturbance.

There are several formats for helical-scan video recorders. *Figure 8.14* shows a complete machine for the Philips 2000 system; *Figure 8.15* shows two internal views of the VHS type.

Video recording on disc

Television recording on disc has the great advantage of low-priced *software* (recorded programme material, films, etc). To copy a programme on to videotape is a relatively slow and exacting business, whereas video discs can be mass-produced by pressing from a master, like their audio counterparts. This price advantage has to be weighed against the consumer's inability to record his own choice of programme material, and it is this factor which, in the eyes of some manufacturers, will limit the popularity of the video disc. Such doubts have not prevented a great deal of development of the disc idea and, as with video tape recording, several formats have been evolved, each quite incompatible with the others. Let us first dismiss any naïve ideas about recording TV pictures on disc. Just as we were disillusioned about extending audio techniques to record TV pictures on tape, so we cannot think of using conventional lateral groove-modulation for recording video on disc. We are not up against octave limitations and dynamic range (though these would pose a problem) so much as the fact that the mechanical nature of the record/replay process, and the physical characteristics of any disc surface, preclude anything like the 3 MHz frequency range necessary for acceptable pictures. Other means of imprinting the picture information on the disc have to be resorted to, and we will briefly describe two systems: optical and capacitive.

Optical system

LaserVision (Video Long-Play), developed by Philips, is the more sophisticated system. It uses a disc similar in dimensions to a standard long-playing audio record. The video information on the disc takes the form of a spiral track of microscopic pits of equal width and depth. The length of the pits and their spacing are determined by the modulating signal – the video waveform – so basically the system is an FM one. Very similar techniques to those used in VCR machines are used in the LaserVision system. Thus we find FM modulation and reduced-bandwidth colour-under in the signal spectrum presented to the disc. The information on the disc is

200

read out by shining a powerful light (sourced from a 1 mW laser) on to the disc surface. Each pit causes a change in the angle of the reflected light from the surface of the disc and thus modulates the light input to a photo-diode which is 'looking' at the track. The ouput of the photo-diode is modulated with information corresponding to the pit length and spacing, and a video signal can be recreated from this. The principle sounds simple enough until it is realised that, to achieve a useful playing time of 30 minutes or one hour per side of disc, the density of information must be very high. The pitch of the pit-spiral is just over $1.5 \mu m$ and each pit is $0.4 \mu m$ in depth and width. At 1500 rpm, this implies over 50 000 revolutions of the disc per side, each revolution containing two television fields, or one complete frame.

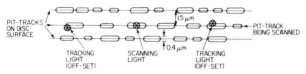

Figure 8.16. The 'tracking lights' of the LaserVision disc system. Two offset tracking beams keep the central scanning beam on the right track through a servo system

The optical system does not involve physical contact with the disc; the laser scanning head 'hovers' just below the disc surface, and the laser beam is brought to an optical focus at the pits on the surface of the disc. This raises the question of tracking – without a groove or any other guide, how do we track the spiral groove accurately, and ensure that the laser beam keeps scanning along the centre of the recorded track? The problem is rather akin to the problem of head-tracking on VCR machines. The main beam, as we have seen, is concentrated on the centre of the pit-track, but two auxiliary beams are also provided, fore and aft of the main beam. These are slightly offset in opposite directions so that they scan partly on and partly alongside the pit track in use, one on each side as shown in *Figure 8.16*. These auxiliary beams are reflected back from the disc into photo-sensors, and if the main beam wanders from the centre of the pit-track, the reflected control beams become unbalanced, one seeing more pits and becoming heavily modulated, the other going 'quiet' as it sees the blank spaces between pit-tracks. This imbalance is detected and used to control a servo-motor driving the laser pickup carriage which re-centres the beam on the track. This all happens necessarily quickly. The carriage travels along a pair of rails beneath the disc. *Figure 8.17* shows a greatly enlarged close-up of the disc surface, and *Figure 8.18* a photograph of the scanning carriage of a LaserVision player.

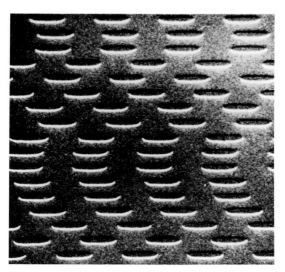

Figure 8.17. A greatly magnified closeup of the pits on the surface of a LaserVision disc (Philips)

The system as described lends itself well to stop-motion and fast or slow forward or backward motion. This is because two television fields are contained in one revolution of the disc, so the field sync pulses all occur on a single line drawn across the diameter of the disc, shown in the diagram in *Figure 8.19*. Each time the readout head crosses this line, we

202

Figure 8.18. The scanning lens of the LaserVision disc player. It is mounted on a carriage which tracks the underside of the disc from the centre outwards. The laser beam passes through the lens on to the disc surface and is reflected back through it on to the signal and servo detectors

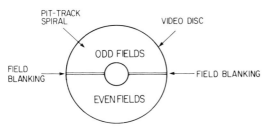

Figure 8.19. Each TV field occupies half a revolution in the 'active' version of the LaserVision disc

have an opportunity to change tracks without picture disturbance, and by 'skipping' the head sideways at this moment, we can re-read one track over and over again to give a disturbance-free still frame. By skipping two tracks backwards on each revolution, we achieve reverse motion; a development of this technique offers half- and double-speed playback.

Because the disc rotates at a constant speed, the outer tracks are longer than the ones nearest the centre of the disc, and this is wasteful in that more information could be packed into the long outer grooves than the single frame as outlined above. In the EP (extended play) version of the disc, the video frames are laid down so that each occupies the same track length regardless of the varying track diameter. In this mode, the rotational speed of the disc varies, being slowest when scanning the outer tracks and fast when the innermost racks are being scanned. The speed control is automatic, and governed by the repetition rate of field sync pulses from the pickup head. This offers a playback time of about one hour, but without the still-frame and 'trick speed' facilities. This is quite acceptable for many applications, such as feature films, and the disc player is designed to cope with either type of disc, automatically selecting the correct replay mode when the disc is inserted.

Capacitive-disc system

An alternative approach to video recording on disc is embodied in the VHD (video high density) system developed by JVC. Again, the picture information is recorded on the disc in the form of a fine spiral of pits in its surface, but here the resemblance to the optical system ends. A relatively large, flat-bottomed stylus is used on the grooveless disc, and the stylus embraces about three pit-tracks at once. At the stern of the sapphire stylus is fitted a metal electrode, and the capacitance between this and the conductive disc changes each time it sees a pit, so that the pulse-coded picture information can be recovered from the rapidly-changing capacitance variations at the stylus electrode.

In addition to the programme pits which carry the picture information, and interlaced with them on the disc, are tracking pits. These are of different shape and periodicity to the programme pits; they are sensed by the pickup electrode, to provide tracking signals for use in servomechanisms to control the position of the stylus arm and keep the stylus tracking down the centre of the programme track. An idea of the stylus tracking is given in *Figure 8.20.*

The cantilever stylus arm is under the control of an electro-magnet system which can move it not only laterally to correct tracking errors, but telescopically, as it were, to compensate for timebase errors. The latter correction has the same effect as momentarily speeding up or slowing down the disc speed, and helps to correct for the line jitter which is inherent in all mechanically-driven picture-replay systems.

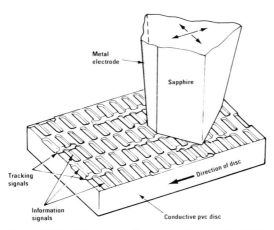

Figure 8.20. The disc surface and stylus of the VHD capacitive disc system. The tracking pits control servos to guide the stylus

The VHD disc rotates rather more slowly than the LaserVision disc. Nominal speed is 900 rpm of a 300 mm PVC disc, offering one hour's playing time per side of the disc. Such facilities as still frame and trick speed are possible with the system. Both VHD and LaserVision systems also lend themselves to high-quality audio recording, as an alternative to the LP sound disc. The VHD system also offers the possibility of replay of either type of disc in the same player. The video disc offers a performance, in its audio guise, which is almost unbelievable to those of us used to the limitations of conventional discs, and such performance parameters as 90 dB

205

signal/noise ratio, no measurable wow or flutter, and ultra-wide frequency response are being suggested.

The future for video recording

As we have progressed through video recording systems in this chapter, we have gradually moved away from the basic analogue video signal which we have studied in previous chapters, and we have seen how the picture signal is recorded on tape in FM form, and on disc in digital form, the information being recorded as a series of pulses with only two states: on (pulse present) and off (pulse absent), corresponding to 1 and 0 respectively. If, then, we can build a large enough memory to hold millions of these 1 and 0 signals, and we are able to read them out in order, we shall have created a static video recording system with no moving parts and none of the complexities of servo systems and mechanical transport of the recording medium. All jitter and wear problems will disappear, and indefinite life can be expected from the system. Memory devices, in silicon chip form, do exist, finding their main use in computer applications. It seems that many years will pass, however, before a memory sufficiently large to accommodate even a 30-minute TV programme will be available. In the next chapter, we shall look at other digital systems related to television.

9

Data transmission and digital TV

To begin to understand TV data displays, we must go right back to the beginning of our study of television, and consider the basic scanning process again. We saw in Chapter 1 that the picture tube is scanned from left to right, and slowly from top to bottom to create a rectangle of light, known as a *raster*. Having set up the raster, we can adjust the intensity of the scanning beam to brighten and darken selected areas of the raster, and thus build up a picture. In a data display, just the same process takes place, but here the video signal has just two states – on (screen alight) and off (screen dark). We no longer have to worry about linearity or dynamic range in the stages handling the video signal, though it is important that such stages have a wide bandwidth (fast rise time) to preseve sharp and crisp edges on the data characters.

Electronic typewriter

Let us imagine that we have set up a raster on the TV screen and have access to a squarewave generator capable of being synchronised to the line timebase frequency of 15.625 kHz. If we connect the output of the squarewave generator to the input of the TV's video amplifier and adjust it to generate one complete square-wave cycle during each $64\,\mu s$ line period, the first half of the scan will be white, corresponding to beam on, and the second half black – beam off. This is about the simplest possible 'pattern'. If we now increase the input frequency so that eight complete squarewave cycles occur during the line period, each complete cycle will take 8 μs (8 × 8 = 64 μs), and in the 52 μs active (picture) line period we shall see 52/8 = 6½ cycles of squarewave, giving rise to seven white vertical bars spaced by six black ones.

We know that one field period contains about 288 'picture' lines and that these are spaced evenly down the screen. If we let our squarewave generator run for say 30 lines then invert its output for the next 30 lines, the positions of the black and white vertical bars will be reversed. Now let the squarewave revert to normal for the next 30 lines, invert for the next 30, and so on, all the way down to the bottom of the screen. The resulting pattern will have 30-line horizontal bands of alternate black and white blocks, and we shall have set up a chequerboard effect on the screen. If we now increase the squarewave frequency, and correspondingly increase the 'inverting' frequency, while keeping them to multiples of line and field rate respectively, we shall generate more and smaller squares in our chequerboard. So by generating short-duration video pulses related to line and field frequencies, we can set up a dot matrix on the display screen, the dots really being small squares of a size determined by the duration of the video pulse. The chequerboard pattern is often used in TV-system testing, and the 'square dot' is familiar to most people as the ball in simple football or tennis TV games.

Having generated a dot matrix, we are free to choose which of the dots appear on the screen by simply opening and closing an electronic gate to let through those we wish to see. We can *program* a gate to let through only those dots which would build up the letter A on the screen, or B, or 6, or whatever. To achieve this with discrete circuits would require a huge number of components, mainly diodes and transistors, but LSI (large scale integration) chips are available which contain all the gates necessary to generate characters or patterns on the screen. Each alphabetic or numerical character has its own gating sequence, and these can be designed into the chip during manufacture, so that by 'enabling' a single pin on the *character-generator* chip, a specific character can be generated and displayed. The gating sequence

208

(unique to each character) determines what we see, and the timing of that sequence, relative to line and field sync pulse, determines its position on the screen. The enabled pulses are, of course, digital (1 to enable, 0 to disable) and may be produced by a simple electric keyboard, with the 'shift' keys altering the timing of the complete characters to move them around the screen.

Let us take away the keyboard again, and think of remotely-generated control pulses. Again, they need be only 0 or 1. If we have, say, 96 characters, we need a digital coding system at the sending end, and a corresponding decoder at the receiver, capable of sending 96 different commands, each finally appearing as a binary digit, either 0 or 1.

Why 96 characters? We need to generate an upper-case alphabet, a lower-case alphabet, numerals, punctuation marks, and also 'blocks' with which to built up simple pictures, known as 'graphics'. Nor is this all. Where the data is 'slotted in' to another information stream, as in teletext systems, we need to send additional information to synchronise the data decoder, to identify page number, time, status of characters (such as colour, flashing, 'concealed'), and so on. A total of 128 'control lines' is used for teletext, for instance. Where the data are being fed into a colour display, some of these commands are used to route the video signals to appropriate guns of the picture tube. By so doing, and still with a binary (two-state) video signal, we can display our data in any of the seven colours represented by white, the three primaries, and their three complementary colours.

Teletext

Teletext is the most familiar form of data transmission, and certainly the cheapest, because it rides 'on the back' of the broadcast TV signal. It is a system for transmitting news and information magazines alongside normal television pictures by sending digital command signals to a suitably-equipped receiver. The receiver uses these commands to type out pages of text or graphics on the TV screen, up to a possible maximum of 800 pages per magazine. For a given number of text lines (currently four), the more pages transmitted, the greater the access time, and to prevent too long a wait between calling up a page and displaying it, current magazines contain 100–200 pages with a worst-case access time of 25 seconds. Control of the teletext receiver is by a keypad, with which the viewer can switch his receiver between normal TV reception and text display, then select the required page of the magazine. The system offers such facilities as subtitling of programmes for the deaf (in which the text is 'boxed' and mixed with the broadcast picture), optionally concealed sections of a page, flashing characters for emphasis, and so on. The service is free to those with a suitably equipped receiver, and one magazine per TV channel is provided by the broadcasters.

Encoding the text signal
Chapter 1 related how the scanning spot is made to fly back to the top of the screen at the end of each field scan, and we saw that a 'settling down period' of several lines is allowed before picture information begins again on line 23 of the first field, and line 336 of the second field. It is during these 'lost' lines that the teletext signals are transmitted, and while as many as 12 lines could be utilised, at present only four lines in each field carry text information. These are lines 15, 16, 17, and 18 in the first field, and corresponding lines 328, 329, 330, and 331 in the second field. These can be examined on an ordinary TV receiver by slipping the field hold or reducing the height, and they appear as rows of twinkling dots.

A text-page display is made up of 23 rows of information, and each row can contain a maximum of 40 characters. Each of the data lines in the transmitted television signal contains sufficient information for one row of characters, so to transmit one 23-row page of text requires six fields or just under one-eighth of a second. We have already seen that the text signal is in binary form; the basic 0 and 1 pulse is called a 'binary digit', 'bit' for short. On each data line, these bits are sent at 444 times line frequency, corresponding to a bit rate of 6.9375 Mbit/sec (1 Mbit/sec = 1 million bits per second).

210

So the 'busiest' data signal, consisting of the series 01010101, etc, would appear as a square wave at 3.4688 MHz.

Each command in the test signal consists of eight bits, and this 'eight-bit bunch' is called a 'byte'. For each character we transmit one byte, consisting of seven digits followed by an *odd-parity bit* whose function is to check the accuracy of the preceeding seven. Possible permutations on seven digits give us 128 combinations, and the code table (simplified) appears in *Figure 9.1*. From this we can see that the byte 1010001 gives capital E, 1110110 gives figure 7, and so on. We cannot start

BITS 7 6 5 4 / 3 2 1	0000, 0001, 0010, 0011	0100	0101	0110	0111	1000	1001	1010	1011	1100	1101	1110	1111
000	THESE CODES ARE FOR CONTROL FUNCTIONS WHICH ARE NOT DIRECTLY DISPLAYED		(	0	8	@	H	P	X	—	h	p	x
001		!	)	1	9	A	I	Q	Y	a	i	q	y
010		"	*	2	:	B	J	R	Z	b	j	r	z
011		£	+	3	;	C	K	S	←	c	k	s	¼
100		$	,	4	(	D	L	T	½	d	l	t	▮▮
101		%	—	5	=	E	M	U	→	e	m	u	¾
110		&	·	6	)	F	N	V	↑	f	n	v	÷
111		'	/	7	?	G	O	W	#	g	o	w	▮

Figure 9.1. Code table for teletext signals

transmitting characters at the beginning of a text line, because the decoder in the receiver needs to be set up and synchronised. We start each line with a two-byte 'run-in' sequence to synchronise the clock oscillator in the decoder, and the next three bytes identify which row of text is being sent. The remaining 40 bytes of the text-data line define the characters in the text-display line being transmitted. In this way, many complete pages of information can be sent over a few seconds. Plainly, though, the text signal will want a lot of sorting out at the decoder.

Teletext decoder

The text decoder currently consists of group of ICs on a printed panel. It takes video from the vision detector output of the set and is linked via an infra-red system to a remote keypad operated by the viewer. Video outputs feed the RGB amplifiers of the receiver, or the video amplifier in a monochrome set. A typical text decoder is shown in skeleton form in *Figure 9.2*.

Initially, we need to gate out the text lines from the rest of the TV signal, and then 'clean up' the data bits to remove the noise and distortion they have picked up during their passage over the air and through the receiver circuits. This is the second block in *Figure 9.2*, and it is followed by the data decoder, which is a form of synchronous detector (we met

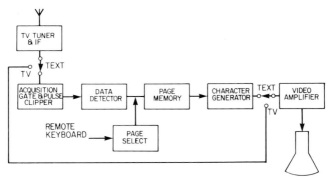

Figure 9.2. Simplified block diagram for a teletext receiver

them in Chapter 4) capable of decoding the eight separate bits in each data byte and presenting them on separate outputs. Let us assume we have selected page 150. This request is keyed into the hand-unit and fed to the decoder, and when the transmitted page number (encoded on to the top row of the page, known as the *header row*) corresponds to the remote command, the page data are fed into the page memory, an LSI chip capable of storing one complete page of text in digital form. Page 150 will then be 'written into' the memory, where it is continuously available even after the transmitted data have moved on to describe other text lines. The page memory must hold 24 lines of 40 characters, corresponding to 960 bytes. This information is constantly read out by the character generator to enable it to build up the data page to normal television standards of 625/50. Thus the page memory is re-read at field rate over and over again until a new page is selected and written into it.

Prestel

An alternative information service is provided by British Telecom in the form of the Prestel service. This has much in common with teletext, and a Prestel display page looks very like a teletext page. Again the page is selected by a keypad and written into a page memory for display. Prestel signals are digitally encoded and reach the receiver via the public telephone network. Because we are not having to slot the data signals into spare moments of a TV transmission, the stream of control information is constantly available once contact has been established, but the transmission path is limited to speech bandwidth, so the very fast data bit rate of teletext is out. The telephone line can reliably accommodate a bit rate of 12 000 bits/sec, and these are sent in the form of audio tones – 2100 Hz for 0, and 1300 MHz for 1. So the data stream consists of alternating bursts of 2100 Hz and 1300 Hz tones, and as such it can be easily recorded on a simple audio tape recorder for later replay as a data display – this is a great advantage of the Prestel system.

We have seen that teletext works by sending a complete row of characters at once in the form of a 45-byte data burst. Because the data link is permanently available, the Prestel system is able to send its data one character at a time and built up the data picture in sequence. The same eight-bit byte is used to describe a character, but a prefix is added in the form of a 'start-bit' and a suffix by a 'stop-bit'. This makes 10 bits per character and implies an eight-second buildup for each page. This would be intolerable where the pages are sent in sequence, because the access time would be much increased. The Prestel system, however, permits access to the information bank by the user, so that only the wanted page is sent down the line, and the eight-second buildup time is the only 'waiting' period. During this time, the display can be seen to be building up the picture by laying down rows of characters as they are recieved.

The viewer's requests to the *data bank* (a large computer, centrally situated) are originated on his keypad and sent back down the line at 75 bits/sec using a two-tone signal similar to that carrying the display data: 450 Hz is used for logic 0 and 390 Hz for logic 1. Let us look at a Prestel system in simple block diagram form – *Figure 9.3*. The receiver (*terminal*) and computer (data bank) are both connected to the telephone line via a *modem* (modulator/demodulator), which converts the digital data and command signals into the two-tone form for transmission over the line. An auto-dialler and auto-answer system act as 'secretary' at each end to establish initial

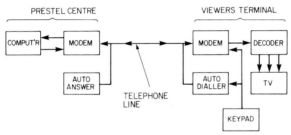

Figure 9.3. Basics of the Prestel system. The transmission medium is the public telephone network

contact between terminal and data bank. To access the Prestel service, the receiver is switched into Prestel mode and the auto-dialler calls up the telephone number of the nearest Prestel centre – the number having been previously programmed into it during installation – whereupon the auto-answer system at the Prestel centre accepts the call and connects the line to the data bank. The computer now sends an index page for display, and by calling up the required page, the viewer can access the information he requires.

The amount of stored information, called the *data base*, is much greater than in broadcast teletext systems, but whereas the latter is free, the Prestel service is operated on a subscription basis. The calling terminal is identified by the Prestel centre at the outset of a call, and the viewer is invoiced in his telephone bill for computer time and for such chargeable pages of data as he may call up; to this is added the standard telephone time-charge for the connection to the Prestel centre.

Prestel receivers for domestic use are also able to receive teletext, and many of the decoder circuits are common to both systems. The talk-back facility built into the Prestel system opens the way for such ideas as electronic mail and, with a security-protected link to the public banks' computers, a complete electronic marketplace, with goods being described, ordered, and paid for without anybody going out of doors.

Digital television

In the last two chapters, we have been discussing digital signals more and more. We have also hinted at some of the advantages conferred by a two-state signal. In our look at videotape and disc systems, we saw that although the signal is binary, the pulse repetition rate is varied to effect modulation, so really they are FM rather than true digital systems. An FM signal needs only a pulse-counter to demodulate or recover the original signal, and a *pulse-width modulated* signal (examples can be found in Chapter 5 in the switch-mode power supply unit and class D field timebase) requires no more than an integrating capacitor – and sometimes an isolating diode – to detect the signal and recreate it in analogue form.

We moved closer to a true digital system in our brief look at data transmission systems. Here we saw that each character is represented by an eight-bit group, or byte, and for the first time encountered the true characteristics of a digital coding system – the idea of equal-length, equally spaced bits, the detection of which requires a synchronous demodulator governed by a synchronised 'clock' oscillator. The data rate for landline transmissions is relatively low thanks to the narrowband telephone network; a much higher data rate is

215

possible with teletext transmissions, but the signal only has a 'window' of, say, eight lines in 625, so it is only available for less than 1½ per cent of the time.

The analogue signal with which we have become so familiar in earlier chapters of this book is very fragile and subject to all sorts of impairements. If it passes through a narrow-bandwidth channel, definition is lost; if it is not amplified in linear fashion, tone-crushing occurs; if it becomes weak or picks up noise, the resulting picture becomes grainy; it is subject to ghosting, phase errors, and impulsive interference. It is degraded when passed through a cable system or a conventional VCR, and whenever it is fed via a capacitor, or any other device which 'loses' the DC component, this has to be restored by clamping circuits. As a rule, once a conventional TV signal has become damaged in any way it is almost impossible to repair it afterwards. Most of the problems associated with generating, recording, propagating, and decoding the television signals stem from this need to preserve the signal intact.

As we said in Chapter 1, the situation may be likened to an attempt to carry a clean white shirt through a working coal mine and present it spotless at the far end. We could wrap our shirt in a bag and thus protect it, but no such 'bag' can be devised for the analogue television signal. Suppose we dispensed with the shirt at the mine entrance, and took instead a detailed description or blueprint of the shirt through the mine? No matter how blackened and crumpled the blueprint became on its journey, provided it were legible, an exact replica of the shirt could be created at the far end. So it is with digital television. If we can devise a system of encoding a full-bandwidth television signal into binary form, then process, record, receive, and decode it as such, many of the old problems will disappear as by magic. The idea is not without problems of its own, however. In an analogue system, the sorts of impairments we have described cause a subjective deterioration in picture quality, and the worse they are, the 'rougher' the picture, or synchronisation, or colour. Picture quality, then, is inversely proportional to the degradation of the signal. In a digital system, the situation

216

is quite otherwise. If the teletext signal is impaired to a point where the inbuilt error correction system cannot cope, the displayed text page will not become noisy or ragged as an analogue-derived signal would – it will start to make errors, and insert incorrect characters, though in perfectly clean and readable form. Thus we may see 'cle n *nh refd£ble ■or#' in place of the last four words of the previous sentence. Such gibberish is the result of the decoder's misinterpreting control commands which have become damaged, and we can see that any digital system is likely to be either perfect or unusable, with no margin between.

There is no chance that the TV transmission system itself could 'go digital' in the forseeable future, because it would render all existing receivers obsolete overnight, as it were. There is, however, much to be said for processing the signal in digital form at the sending end, and in the domestic receiver. Let us look at some of the advantages for the broadcaster. When the television signal is in digital form, it can be constantly written into a field store, which is a larger version of the page memory we met in the teletext decoder. The broadcaster can read out of the field store at a different scanning phase, or rate, so the necessity to 'slave' or *genlock* signal sources together for mixing would disappear. Television standards converters make use of this technique, and freeze-frame can be instantly achieved by stopping the writing process. Digital videotape recording provides a playback signal indistinguishable from the original material, even after many *generations* or re-copying processes. Electronic circuits for dealing with signals of only two levels are simple to design and cheap to manufacture. Digital signals lend themselves to electronic 'effect' processing easily, and pictures can be inverted, mirrored, 'squeezed', and zoomed if desired, by altering the memory readout direction and speed. The ultimate goal is the all-digital studio, where the picture exists in analogue form only on the camera faceplate, and many 'pictures', or at least their components, are artificially generated within digital chips. . . Let us come back to earth and see how an analogue signal is 'digitised'.

217

Analogue-to-digital conversion

To convert an analogue signal to digital form, it is necessary to sample its level at fixed intervals and assign each level a number. Consider the waveform in *Figure 9.4*. Let us suppose that a reasonable facsimile of it can be achieved by transmitting 25 levels of amplitude. If it has a period of one second, we may choose to sample its level at 100 ms intervals, which will give 10 samples of its amplitude. Reading from the diagram, the sampling levels are: 5, 11, 16, 19, 22, 22, 16, 15, 13, 3, 5. These are whole numbers, and each represents the

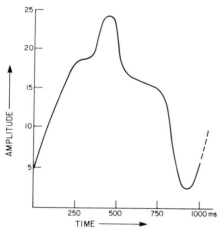

Figure 9.4. A complex analogue waveform

BITS 5		0	1	0	1	0	1	0	1
4		0	0	1	1	0	0	1	1
3		0	0	0	0	1	1	1	1
0	0	0	1	2	3	4	5	6	7
1	0	8	9	10	11	12	13	14	15
0	1	16	17	18	19	20	21	22	23
1	1	24	25	26	27	28	29	30	31

Figure 9.5. Five-bit code table offering 32 commands

218

instantaneous voltage of the waveform at the moment of sampling. Each number can be assigned a unique bit-combination, so, referring to the table in *Figure 9.5*, we can see that our waveform would appear as 00101, 01011, 10000, 10011, 10110, 10110, 10000, 01111, 01101, 00011, 00101. We have laid the digits out in this way to indicate the considerable string of bits necessary to convey even a simple waveform, crudely analysed. Let us see what happens when we decode our digital signal and reassemble it in analogue form. *Figure 9.6* shows the result. It is a copy of sorts and may do

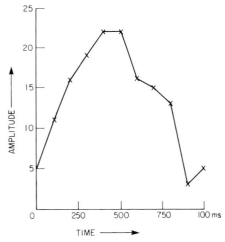

Figure 9.6. Recreated analogue waveform
– compare with *Figure 9.4*

for some applications; it certainly would not do for television. We have lost the 'shoulder' at 300 ms altogether, and the whole waveform is somewhat rounded off.

The answer to these problems is to sample more frequently and define more levels. In fact a sampling rate of at least twice the highest frequency of interest is necessary; for 625/50 standard television, whose bandwidth extends up to 5.5 MHz, we must sample at 11 MHz rate or faster. In our simple example, we chose 25 levels of amplitude and lost some detail thereby. We were, however, able to accommodate our 25 levels, and potentially more, in a five-bit 'word'. It has been shown that 25 levels of grey would give a totally unacceptable picture, but if we can cater for 200 shades of grey, the resulting picture is indistinguishable from the original. An eight-bit word, or byte, can accommodate 256 levels, and this is the standard used in digital television. The luminance information is sampled at 13½ MHz rate and split into 256 levels. To give a little headroom, black is assigned level 16 and peak white level 235. We are going to need a high bit-rate at this sort of sampling frequency, and we have done nothing about chrominance signals yet. The colour-difference signals are sampled at half the rate for the luminance signal: i.e., 6½ MHz for each colour-difference signal, B-Y and R-Y. These signals, unlike the luminance component, can have positive or negative values, so with the same 256-level capability, zero is represented by level 128 with positive-going signals having levels up to 239 and negative-going signals occupying levels 128 down to 16. When the digital signals for chrominance and luminance are combined for transmission, the bit-rate comes out 216×10^6/sec, or 216 Mbit/sec. Such is the price of full-definition colour TV.

Digital signal processing at the receiver has much to offer. Gone will be internal pre-sets, delay lines, and most wound components. Flutter and noise will disappear, and with them any disturbing effect on timebase synchronisation. Reception of two broadcast channels simultaneously will be possible, with the 'guest' picture in one corner of the 'host' picture. Zoom effects and still-frame pictures will be possible at the touch of a button, and by reading out of a digital field store at, say, 75 fields/sec instead of the usual 50, we can eliminate the slight flicker inherent in the current system when displaying pictures at high brightness levels. The hardware to carry out this signal processing comes in the form of VLSI (very large-scale integration) chips. The first system intended for receiver applications was demonstrated in 1981 by ITT Semiconductors Group. We've come a long way from the bulbs and photocells in Chapter 1!

Further reading

The following books, all published by Newnes Technical books, will provide further information on various aspects of radio, television and video that cannot be covered in greater detail in this book.

Beginner's Guide to Electronics, 4 ed, Owen Bishop
Beginner's Guide to Integrated Circuits, Ian R. Sinclair
Beginner's Guide to Radio, 8 ed, Gordon J. King
Beginner's Guide to Transistors, 2 ed, Ian R. Sinclair and J. A. Reddihough
Beginner's Guide to Video, David K. Matthewson
Colour Television Servicing, 2 ed, Gordon J. King
Electronics Pocket Book, 4 ed, E. A. Parr
Foundations of Wireless and Electronics, 9 ed, M. G. Scroggie
Home Video, Richard Dean
Integrated Circuit Pocket Book, R. G. Hibberd
Newnes Book of Video, ed, K. G. Jackson
Newnes Colour Television Servicing Manual, Vols. 1–3, Gordon J. King
Principles of PAL Colour Television, H. V. Sims
Questions and Answers on Colour Television, 3 ed, E. Trundle
Questions and Answers on Electronics, 2 ed, Ian Hickman
Questions and Answers on Integrated Circuits, 2 ed, R. G. Hibberd
Questions and Answers on Radio and Television, 4 ed, H. W. Hellyer and Ian R. Sinclair

Index

223

225